CAMBRIDGE MONOGRAPHS ON PHYSICS

GENERAL EDITORS

N. FEATHER, F.R.S.
Professor of Natural Philosophy in the University of Edinburgh

D. SHOENBERG, PH.D.
Fellow of Gonville and Caius College, Cambridge

THE ADSORPTION OF GASES
ON SOLIDS

THE
ADSORPTION OF GASES
ON SOLIDS

BY

A. R. MILLER

*Imperial Chemical Industries Research Fellow
at the Royal Society Mond Laboratory,
Cavendish Laboratory, Cambridge*

*A sequel to the Cambridge Physical Tract
written in 1939 by J. K. Roberts and entitled
Some Problems in Adsorption*

CAMBRIDGE

AT THE UNIVERSITY PRESS

1949

CAMBRIDGE UNIVERSITY PRESS
Cambridge, New York, Melbourne, Madrid, Cape Town,
Singapore, São Paulo, Delhi, Mexico City

Cambridge University Press
The Edinburgh Building, Cambridge CB2 8RU, UK

Published in the United States of America by Cambridge University Press, New York

www.cambridge.org
Information on this title: www.cambridge.org/9781107621428

© Cambridge University Press 1949

First published 1949
First paperback edition 2013

A catalogue record for this publication is available from the British Library

ISBN 978-1-107-62142-8 Paperback

GENERAL PREFACE

The Cambridge Physical Tracts, out of which this series of Monographs has developed, were planned and originally published in a period when book production was a fairly rapid process. Unfortunately, that is no longer so, and to meet the new situation a change of title and a slight change of emphasis have been decided on. The major aim of the series will still be the presentation of the results of recent research, but individual volumes will be somewhat more substantial, and more comprehensive in scope, than were the volumes of the older series. This will be true, in many cases, of new editions of the Tracts, as these are republished in the expanded series, and it will be true in most cases of the Monographs which have been written since the War or are still to be written.

The aim will be that the series as a whole shall remain representative of the entire field of pure physics, but it will occasion no surprise if, during the next few years, the subject of nuclear physics claims a large share of attention. Only in this way can justice be done to the enormous advances in this field of research over the War years.

N. F.
D. S.

CONTENTS

CHAPTER 6

Some Other Types of Adsorption

CHAPTER 7

Dipole Interactions between Adsorbed Particles

AUTHOR'S PREFACE

Kinetic theory calculations can be carried out exactly when the dimensions of the apparatus used are small compared with the free path of the molecules in the gas. The discrepancies between such calculations and experiments on the exchange of energy between a metal surface and a gas at a different temperature suggested that the metal surfaces which had been used were all covered with adsorbed layers of impurity. To obtain results which would refer to known systems, it was necessary to work with surfaces kept free from impurities. This formed the starting-point for the experiments initiated by the late J. K. Roberts in 1931. These experiments, together with the theory so far as it had been developed by the end of 1938, were described in a Cambridge Physical Tract entitled *Some Problems in Adsorption*. The series of physical tracts were designed to provide interim reports of research work in progress, and it was intended that they should be replaced, as the development of their subject-matter demanded, by revised editions which would contain accounts of subsequent research. The present monograph is such a revision of Roberts's tract on adsorption. When the original tract was written statistical theories had not been developed to describe either the behaviour of immobile films (which had been studied by empirical methods) or of the effect of dipole-dipole, in addition to van der Waals, interactions; nor had account been taken of the continuous variation of the potential field provided by the adsorbing surface. In considering a revision of the original tract it seemed desirable to rewrite it so that these matters could be dealt with adequately. To do this it was necessary to discard much of the original text and to rearrange the presentation of the remainder to conform to the new material which was to be introduced. In the result, about half of the present book consists of new material, which is contained mainly in Chapters 2, 3 and 7, with important additions also to Chapters 4 and 5.

A. R. MILLER

Royal Society Mond Laboratory
University of Cambridge
21 MAY 1947

CHAPTER I

EXPERIMENTAL METHODS

1·1. General Introduction

The classical experiments of Langmuir showed that, when a clean surface of a solid or a liquid is exposed to an impurity, a monomolecular layer of the impurity is often formed on the surface and contaminates it. In the study both of the behaviour of surfaces and of surface reactions it is of considerable importance to understand the properties of such monomolecular films. Furthermore, many properties of surfaces, for example, the interchange of energy between a solid surface and a gas,* depend in a very noticeable way on the presence of gas atoms or molecules adsorbed on the surface. To understand the processes which take place at the surface it is necessary to study the properties of such films and the physical mechanism by which they are formed. It is the purpose of this monograph to consider some of the experimental results concerning films of gases and vapours adsorbed on solid surfaces and to present the statistical thermodynamical theory in terms of which the properties of these films can be described.

It was first realized by Knudsen (1911, 1915) that, when the relevant dimensions of the apparatus used are small compared with the free path of the molecules, calculations based on the kinetic theory of gases can be carried out exactly. He performed a number of very beautiful experiments which substantiated this view, thus making a great advance in the theory, since phenomena occurring in the gas phase were completely elucidated by his work.

In dealing with the exchange of energy between a solid and a gas at a different temperature, Knudsen introduced a quantity called the accommodation coefficient. It is a measure of the extent to which the gas molecules leaving the solid are in thermal equilibrium with it and is defined in the following way. Gas atoms at temperature T_1 strike a solid surface which is at temperature T_2

---

* This is discussed further below.

(see fig. 1). On leaving the surface, the gas atoms will not in general be in equilibrium with it, i.e. they will not be at the temperature T_2, but will on the average have energy corresponding to a temperature T_2'. The accommodation coefficient a is defined by the equation

$$a = \frac{T_2' - T_1}{T_2 - T_1}. \tag{1·1}$$

The details of the measurement of a and the proof that to a first approximation it is a constant, will be considered below. For a equal to unity the gas molecules leaving the surface are in equilibrium with it, while for a zero there is no net interchange of

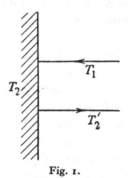

Fig. 1.

energy between the gas and the solid. Experimental determinations of a for various gases were made by Knudsen and other workers. On the basis of suitable simplifying assumptions, Baule worked out a classical theory of the collisions of the gas molecules with the atoms of the solid. This gave an expression for the accommodation coefficient of a given monatomic gas and a given solid which contained no arbitrary constant. The values so calculated were consistently very much smaller than the measured values. Further, while the theory indicated that the accommodation coefficient should depend on the ratio of the masses of the gas and solid atoms, the experiments showed that for a given gas the accommodation coefficient under comparable conditions was nearly the same for all the metals tried.

These two results together suggested that the metals were all covered with adsorbed layers of impurity of unknown composition, so that the application of a physical theory was impossible.

To obtain experimental results that would be useful in testing any theory, it was essential to work under such conditions that the adsorbed layers could be removed and the surface kept free from impurities. To carry out experiments under these conditions formed the starting-point of the experiments initiated by Roberts in 1931. The experiments were extended to a wide range of interactions between gases and solids and to the study of the structure and behaviour of adsorbed films of gases built up on a known substrate. Their purpose was to obtain experimental data which should provide the basis for a physical theory of the interaction between gas atoms and solids. It should, perhaps, be noted here that there is no need to discuss Baule's simple classical theory of collisions, for it was quickly superseded by a quantal treatment.

It was essential to obtain experimental results under such conditions that the natures of the gas and solid atoms concerned in any interaction were known. There is little difficulty in this as far as the gas is concerned, but with the solid special care must be taken. In general it is best to work with a metal, so that all the atoms are of the same nature. At the beginning of the experiment the metal must be freed from all adsorbed films of impurities, and precautions must be taken to remove all adsorbable impurities from the gas so that the surface remains clean during the course of the experiment. In addition to the fact that the gas and solid atoms are known, experiments carried out under such conditions have the advantage that the system is a comparatively simple one. On the other hand, a surface covered with adsorbed films of impurity of unknown structure and composition is a highly complicated system, and it would be expected on general grounds that results of a much more fundamental character would be obtained when working with the simple than with the composite system.

Apart from this particular application of it, the idea of starting with simple systems was essential, for only in this way was it possible to build up a detailed picture of what was happening. We are concerned with the phenomena which occur when a gas atom approaches a solid surface. It may either make a collision and rebound, an interchange of energy in general taking place, or else it may remain an appreciable time on the surface, i.e. be adsorbed. The behaviour would be expected to be simplest with helium, for here the attractive forces between gas atoms and the surface are very weak, so that adsorption effects would be practically absent except at very low temperatures, and collisions with the accompanying interchange of energy would alone be involved. In his first experiments Roberts studied the interchange of energy between helium and a tungsten surface by measuring the accommodation coefficient.

In previous experiments by various workers in which an ordinary wire was used, it was found that the accommodation coefficient of helium at room temperature was always about 0·3. When such

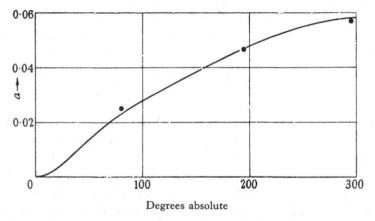

Fig. 2. Variation of accommodation coefficient of helium with temperature.

precautions are taken as are necessary to ensure that the surface of the tungsten is bare there is a striking change, and a value in the neighbourhood of 0·05 or 0·06 is obtained (Roberts 1932 a). Not only is the actual magnitude of the accommodation coefficient affected, but the nature of its variation with temperature is quite different. Soddy and Berry (1911) found for helium with ordinary platinum wire a value of 0·28 in the neighbourhood of 20° C. and of 0·36 in the neighbourhood of −185° C. With bare tungsten, on the other hand, the temperature variation is shown by the points in fig. 2. The general run of these points suggests that the accommodation coefficient approaches zero as the absolute zero is approached, but no doubt the lowest parts of the curve would not be realized in practice owing to condensation of the helium on the surface.

The following (Roberts 1932 b) physical picture explains this falling off at low temperatures. To begin with, let us regard the solid as an assembly of Planck oscillators all of identical frequency v, as in Einstein's original theory of the temperature variation of the specific heat of solids. When a gas atom interacts with such an assembly, energy transfers can take place only in amounts of nhv, where n can have the values 0, 1, 2, 3, etc. At temperatures at which the mean thermal energy is smaller than hv a considerable number of the oscillators will be in the ground state. When a gas atom interacts with an oscillator in the ground state, the only possible interchange of energy is that in which a gas atom gives up energy to the oscillator, and the smallest energy that the oscillator can take is hv. The number of gas atoms that have energy hv to give up becomes progressively smaller as the temperature gets lower, and thus the proportion of gas atoms which can undergo a change of energy on interacting with the solid becomes smaller and smaller as the temperature approaches the absolute zero. If the solid is not regarded as an assembly of oscillators all having identical frequencies but having frequencies distributed over a range from 0 to v_m, the above considerations show that at sufficiently low temperatures only the lower frequency oscillators will be appreciably affected by collisions with gas atoms. Therefore, fewer and fewer oscillators are affected as the temperature falls. As a result, the interchange of energy, under comparable conditions, becomes less and less efficient. Thus for a small temperature difference between solid and gas the accommodation coefficient becomes smaller and smaller.

Devonshire (1937), in one of a series of papers by Lennard-Jones and his co-workers, has followed up the work of Jackson, Mott and others (Jackson 1932; Jackson and Mott 1932; Jackson and Howarth 1933, 1935; Landau 1935) in developing the detailed quantum theory of collisions of gas atoms with solids. The efficiency of energy interchange between the gas atom and the oscillations of the solid depends, among other things, markedly on the masses of gas and solid atoms, which are known, and on the law of force between the gas atom and the atom of the solid. Devonshire assumed that the potential energy between a gas atom and an atom in the surface of the solid when their centres are at a distance z apart is given by a Morse function

$$D\,e^{-2\kappa(z-b)} - 2D\,e^{-\kappa(z-b)}.$$

This is plotted in fig. 3, in which the meanings of b and of D are shown; D is closely related to the heat of adsorption. The third constant κ determines the rate at which the repulsive potential energy increases when the atoms are closer together than b. The greater κ is, the more sharply does the repulsive potential increase and, other things being equal, the greater is the accommodation coefficient. For helium, D is small and the curve in fig. 2 is the theoretical curve for a one-dimensional model for $\kappa = 2 \times 10^8$ and $D = 0$ multiplied throughout by a roughness factor $1 \cdot 06$* to make it fit the experimental point at $195°$ K.

As has been mentioned, the behaviour with helium would be expected to be simpler than with any other gas because of the small value of D (fig. 3). The next gas used was neon, for which D is appreciable. This makes the curve showing the variation of a with temperature fall off less slowly than it would if D were zero, and Devonshire (1937) has shown that an accurate experimental determination of the shape of this curve would make it possible to deduce the values of the constants in the Morse function. The experiments described here have determined only the relative behaviour of helium and neon but have not given the shape of the neon curve with sufficient accuracy for the above purpose. The important point about the results for neon from the present point of view is the very large difference between the accommodation coefficient for a surface covered with adsorbed films (about $0 \cdot 6$) and for a clean surface (about $0 \cdot 07$). This suggests that the accommodation coefficient of neon is very sensitive to the presence

---

* Actual surfaces are never absolutely plane. The effect of any roughness is to increase the measured accommodation coefficient over that for a smooth surface, because a certain number of the gas molecules after making a collision with and leaving one part of the surface will strike it again in another place. Thus to obtain the accommodation coefficient for a smooth surface, which is what is required for comparison with the theory, the measured values at various temperatures must all be reduced by dividing by a roughness factor. The value of this factor can be determined only by direct observation of the shape and size of the irregularities on the surface. For a general discussion of this point see Roberts (1930).

of adsorbed films on the surface, and that therefore it can be used satisfactorily as an indicator in studying the adsorption of gases on bare tungsten.

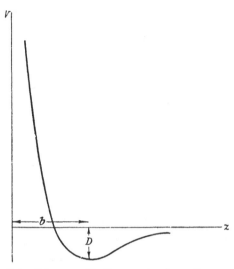

Fig. 3. Potential energy V of two molecules a distance z apart showing the meanings of D and b in the Morse function.

1·2. The Measurement of Accommodation Coefficients

We need consider only the accommodation coefficients of monatomic gases and are therefore not concerned with any difficulties that may arise if the accommodation coefficient for rotational energy is not the same as that for translational energy.*

Suppose that a fine wire maintained at temperature T_2 by an electric current is stretched down the centre of a cylindrical glass tube at temperature T_1 containing a monatomic gas at such a low pressure p dynes cm.$^{-2}$, that the mean free path of the gas atoms is long compared with the radius of the tube and that the diameter of the wire is small compared with that of the tube. Under these conditions any molecule leaving the wire makes many collisions with the tube before returning to the wire again, and it may be assumed that the molecules striking the wire are at the temperature T_1 of the tube. Suppose there are $f(c)\,dc$ molecules per cm.^{3} with velocities lying between c and $c+dc$ cm. per sec. The

* For a discussion of this point see Rowley and Bonhoeffer (1933).

number of such molecules of mass m striking unit area of the wire per second is $\frac{1}{4}f(c)\,c\,dc$, and the energy brought up to unit area of the wire per second by such molecules is $\frac{1}{8}mf(c)c^3dc$. The total energy brought up to unit area of the wire per second is

$$\frac{1}{8}m\int_0^\infty f(c)\,c^3\,dc.$$

Since
$$f(c) = \frac{4n}{\sqrt{\pi}}\left(\frac{m}{2kT_1}\right)^{\frac{3}{2}} e^{-mc^2/2kT_1}\,c^2,$$

where n is the number of atoms per cm.^{3}, the total energy brought up to unit area of the wire per second is

$$n\left(\frac{2k^3T_1^3}{\pi m}\right)^{\frac{1}{2}},$$

and the total number of molecules striking unit area of the wire per second is
$$n\left(\frac{kT_1}{2\pi m}\right)^{\frac{1}{2}} = \frac{p}{(2\pi mkT_1)^{\frac{1}{2}}},$$

where p dynes cm.$^{-2}$ is the pressure. The expression for the energy brought up to unit area of the wire per second may be written

$$\frac{p}{(2\pi mkT_1)^{\frac{1}{2}}}\,2kT_1.$$

Let $n_c dc$ be the number of molecules with velocities between c and $c + dc$ leaving unit area of the wire per second. These molecules are not in thermal equilibrium with the wire at temperature T_2 but have mean energy corresponding to a temperature T_2'. Their distribution of velocities will not therefore be exactly Maxwellian, but the departure from such a distribution becomes smaller as $T_2 - T_1$ becomes smaller.[*] We assume a Maxwellian distribution so that $n_c = Ac^3\, e^{-mc^2/2kT_2'}.$

The total number of molecules leaving unit area of the wire per second is
$$\int_0^\infty n_c\,dc = A\,\frac{2k^2T_2'^2}{m^2}.$$

In the steady state the number of molecules leaving unit area of the wire per second must equal the number striking it so that

$$A\,\frac{2k^2T_2'^2}{m^2} = \frac{p}{(2\pi mkT_1)^{\frac{1}{2}}}.$$

[*] For a discussion of this see Knudsen (1911, 1915).

This condition determines the constant A. The energy carried away from unit area of the wire per second is

$$\frac{Am}{2}\int_0^\infty c^5 e^{-mc^2/2kT_2'}dc = 4A\frac{k^3T_2'^3}{m^2} = \frac{p}{(2\pi mkT_1)^{\frac{1}{2}}}2kT_2'.$$

The net energy loss in ergs per unit area per second is

$$\frac{p}{(2\pi mkT_1)^{\frac{1}{2}}}2k(T_2'-T_1),$$

where k is measured in ergs. Using equation (1·1), this becomes

$$\frac{ap}{(2\pi mkT_1)^{\frac{1}{2}}}2k(T_2-T_1) \quad \text{ergs per cm.}^2 \text{ per sec.,}$$

or

$$\frac{7\cdot3 \times 10^3\, ap(T_2-T_1)}{(MT_1)^{\frac{1}{2}}} \quad \text{ergs per cm.}^2 \text{ per sec.,}$$

where M is the molecular weight referred to $O_2 = 32$ and p is the pressure in dynes cm.$^{-2}$. The heat loss q in calories cm.$^{-2}$ per sec. is therefore given by

$$q = 1\cdot74 \times 10^{-4}\frac{ap}{\sqrt{(MT_1)}}(T_2-T_1). \tag{1·2}$$

All the quantities in this equation except a can be measured and thus a is determined. The wire was placed in one arm of a Wheatstone bridge. q was obtained by measuring the current through the bridge, and the temperature excess of the wire $T_2 - T_1$ is found from its resistance. The pressure was measured on a Macleod gauge. It must be shown experimentally that the value of a obtained does not depend on p or on $T_2 - T_1$. The fact that this last condition is fulfilled establishes the correctness of the assumptions on which equation (1·1) is based.

Actually it is not necessary to work at such low pressures that the free path l is long compared with the radius of the containing tube, but it is sufficient that l should be long compared with the diameter of the wire. Molecules leaving the wire travel, on the average, a distance l before making a collision and then make many collisions with other gas molecules before returning to the wire. The passage of the heat from a circle of radius l to the outside of the tube may be treated as ordinary gas conduction. Simple numerical calculations show that the temperature drop in the gas is small compared with the temperature difference between the wire and the gas at a distance l from it.* Thus the molecules striking the wire can be taken as being at the temperature T_1 of the outer tube. The above theory is the same as that first given and tested by Knudsen (1911, 1915).

In order to be able to measure the accommodation coefficient for a bare surface and to study the adsorption of known gases on

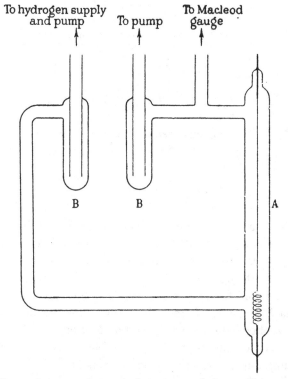

Fig. 4. Apparatus for measuring accommodation coefficients.

such a surface it is necessary to work under carefully controlled conditions. The essential parts of the apparatus that were used for this purpose are shown in fig. 4. The tungsten wire of diameter 0·0066 cm. and about 18 cm. long was contained in the tube A. The neon was continuously circulated by a diffusion pump through the charcoal tubes B immersed in liquid air to remove from it any adsorbable impurities coming from the glass. There were liquid-air

* See, for example, Blodgett and Langmuir (1932).

traps between the charcoal tubes and the rest of the apparatus and two liquid-air traps between the tube containing the wire and the Macleod gauge, and the usual precautions necessary for obtaining the best vacuum were taken.

After the neon at a pressure of about o·1 mm. of mercury had circulated for some time, the wire, which had previously been thoroughly outgassed, was heated to a temperature well above 2000° K. so as to remove all adsorbed impurities from its surface. It was connected to a Wheatstone bridge and, when it had cooled down, a measured current was passed through it sufficient to raise its temperature about 20° above that of the oil bath in which the containing tube was immersed. Taking the time of starting the experiment as the moment when the flashing current was cut off, readings of resistance as a function of the time were taken and the pressure was read on the Macleod gauge. After the temperature coefficient of resistance for the wire had been determined it was possible to deduce from each reading a value of the accommodation coefficient.* These values are plotted in the first part of the curves of fig. 5 which refer to experiments at 295 and 79° K. They show a very slow drift with time, presumably due to the adsorption on the wire of very small residual traces of impurity in the neon. Extrapolation to zero time gives the accommodation coefficient for a clean wire.† At the time indicated by the arrow one dose of hydrogen contained between two taps forming a gas pipette and sufficient to produce a pressure of about 10^{-4} mm. of mercury in the apparatus was admitted so that it had to pass through a charcoal tube before reaching the wire. The accommodation coefficient rose rapidly to a final steady value. This indicates adsorption of the hydrogen on the tungsten. The final steady value was not affected if very much larger doses of hydrogen were admitted, which suggests that a complete monomolecular layer is formed even when the hydrogen is present at a partial pressure of only 10^{-4} mm. of mercury, and this view is completely confirmed by the experiments of a different kind described in the next section. To show that the effects observed were not due to the adsorption of traces of oxygen in the hydrogen, the experiments were

---

* For the correction for radiation and end losses see Roberts (1932a).
† See Roberts (1935a), footnote on p. 447, and Michels (1937).

carried out admitting, instead of hydrogen, the same volume of air at the same pressure as the hydrogen. No trace of the phenomena described above was observed; this means that the charcoal tubes were able to remove completely this relatively very large amount of oxygen.

Blodgett and Langmuir (1932) had previously measured the accommodation coefficient of hydrogen with a tungsten surface.

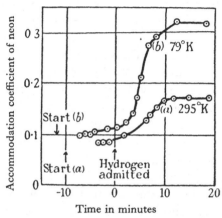

Fig. 5. Effect, on the accommodation coefficient of neon, of admitting a trace of hydrogen when the tungsten surface is clean.

Their experiments showed certain changes in the accommodation coefficient according to the heat treatment of the wire, and they interpreted these as due to changes in an adsorbed film of hydrogen on the surface. The present experiments show conclusively that their view that a hydrogen film is formed on a bare tungsten surface is correct.

The rapid adsorption of hydrogen on tungsten occurs only when the surface of the metal is bare, and earlier workers,* using tungsten powder and the usual methods of cleaning such powders, did not observe it, from which it must be concluded that these methods do not in fact produce a bare metal surface at all. In this connexion some results obtained by Burdon (1935, 1940) are of interest. He has shown that, when a new surface of mercury is formed in the presence of hydrogen at a pressure of 400 mm. of mercury, a

---

* For example, Frankenburger and Hodler (1932). For a discussion of the earlier work see Roberts (1935 a).

monomolecular film of the gas is adsorbed on the surface and held tenaciously, the rate of loss being inappreciable when the pressure is reduced to 2×10^{-4} mm. If, on the other hand, the new surface is first formed in a vacuum and hydrogen then admitted, no such adsorption occurs, presumably because the surface is immediately contaminated. It is a striking fact that this rapid adsorption of hydrogen occurs on two metals so different as tungsten and mercury provided the proper precautions are taken to expose a bare surface to the gas, but that in the absence of the necessary precautions no appreciable adsorption is observed in either case.

It is important to investigate as fully as possible the properties of the hydrogen film and films of other simple gases on tungsten and any other metal that can be cleaned satisfactorily. An understanding of these simplest systems is fundamental to our knowledge of the nature of adsorption processes in general.

1·3. The Measurement of the Heat of Adsorption

The experiments on the accommodation coefficient indicate that a complete monomolecular film is formed at a very low pressure of hydrogen. This, and the fact that the adsorption takes place rapidly, makes it possible to measure the heat of adsorption on a single fine wire. It is, of course, necessary to use a wire or fine strip if it is to be cleaned before the hydrogen is admitted to it.

The idea underlying the method was that the wire on which the adsorption took place should itself form the calorimeter. There is a great advantage in having the heat liberated just where it is wanted, and in knowing the apparent area of the surface on which adsorption takes place. The thermal capacity of the wire is known from its diameter and length (in the actual experiment these were 0·0066 and 28·2 cm. respectively), and the density and specific heat of the material of which it is made are also known. If its mean rise of temperature when the adsorption takes place can be measured, the total heat given out by the adsorption process is obtained.

The apparatus (Roberts 1935 a) is shown in fig. 6. The wire was contained in a tube W protected from mercury and other vapours by liquid-air traps. This tube, and the tube G containing a Pirani gauge, were immersed in a large Dewar vessel filled with oil. The necessary small amount of hydrogen was obtained by expanding one charge of the gas pipette P, which was connected to the hydrogen reservoir, into a large volume (not shown), and then admitting to the part of the apparatus containing the wire the amount of this expanded gas contained in the small volume A between the two mercury cut-offs.

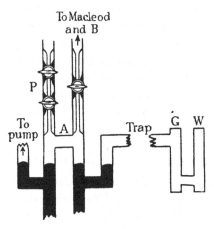

Fig. 6. Apparatus for measuring heat of adsorption of hydrogen.

The actual experimental procedure in its simplest form was as follows:

(i) The wire was flashed and, when the flashing current was cut off, it was connected to a sensitive bridge in which a Paschen galvanometer was used. As the wire cooled down the galvanometer showed a drift with time. After about 10 minutes this drift was sufficiently slow to be easily followed and controlled by altering one of the other arms of the bridge. When this stage was reached the apparatus was cut off from the pumps, the hydrogen charge was prepared, and regular readings of the galvanometer deflexion and time were taken. The hydrogen in A was then admitted to the apparatus, and the galvanometer showed a deflexion in the direction corresponding to a rise in the temperature of the filament as shown in fig. 7. The time of admission, the maximum reading, and the time of its occurrence were noted. The deflexion of the Pirani gauge was read at the same time. The hydrogen was now pumped out, and the deflexion of the Pirani gauge measured as a check.

(ii) The volume A was then refilled so that the pressure in it was practically the same as in (i). This hydrogen was admitted to the evacuated apparatus without flashing the wire. The deflexion of the Pirani gauge was measured and was greater than for the same amount of hydrogen in (i) because there was no adsorp-

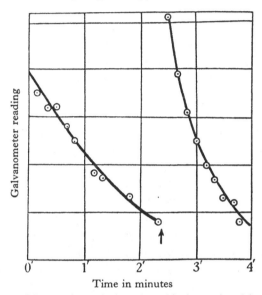

Fig. 7. Effect on heat of adsorption of hydrogen in raising the temperature of a bare tungsten filament.

tion on the unflashed wire. At the same time it was observed, as would be expected, that there was practically no deflexion of the Paschen galvanometer. The Pirani gauge was calibrated by diminishing the bridge sensitivity in a known ratio and comparing the gauge directly with a Macleod gauge.

Provided the volume of the apparatus is known, the readings of the Pirani gauge give the amount of gas adsorbed on the wire in (i) directly. The volume was determined in the ordinary way by expanding air at a few cm. pressure from a measured volume into the apparatus, and measuring the pressure change.

To deduce from the readings the rise in temperature of the filament, a correction had to be made for the cooling during the time when the deflexion was occurring. This correction, which was small and never more than about 10 % of the total deflexion, was obtained in the usual way by a study of the slope of the temperature-time curve at different points. The change of resistance, corresponding to a given deflexion, was obtained by altering by a small known amount a large shunt in the opposite arm of the bridge, the resistance of which was equal to that of the filament. The temperature change was deduced from the temperature coefficient of resistance of the tungsten as determined in an independent experiment. The heat evolved in the adsorption process is given by $ms\Delta T$, where m is the mass of the wire, s the specific heat, and ΔT the rise of temperature.

The above description gives the principle of an experiment which determines the average heat of adsorption when the surface was fully covered at the end of the experiment. This procedure was not followed exactly, but the amount of hydrogen admitted in one charge of A was made less than was necessary to cover the surface completely. When the first charge was admitted there was a heating effect, but there was no pressure change sufficient to be indicated on the Pirani gauge (that is, it was less than 2 or 3 × 10^{-7} mm.). Thus even at these low pressures adsorption was complete and rapid. Charge after charge was admitted, and the heating effect measured each time. Finally, when saturation was reached, the Pirani gauge showed a deflexion, and from this it was possible to deduce the amount admitted at each charge and so to obtain the variation of the heat of adsorption as the surface became progressively more covered.

1·4. The Amount of Hydrogen Adsorbed on Tungsten

We shall consider the number of molecules adsorbed and the number of tungsten atoms in the surface. It has been usual, following Langmuir (1932), to assume that with aged tungsten the 110 plane is the important one in the surface, but R. P. Johnson (1938) has shown that this is not necessarily correct and that the 100 plane must also be considered. Tungsten is a body-centred cubic lattice, and the arrangements of the atoms in these two planes are shown in fig. 8. The numbers of atoms per cm.^{2} are 14·24 × 10^{14} for the 110 plane and 10·07 × 10^{14} for the 100 plane. During the flashing of the filament used in the experiment described in the preceding section, short lengths at the ends would be below 2000° K. These would not be completely freed from any adsorbed oxygen and were therefore not available for the adsorption of hydrogen. Allowing for this the effective apparent superficial area of the filament was 0·55 cm.[2]. Thus, if the wire were smooth, the number of tungsten atoms in the surface would be $7·8 \times 10^{14}$ for the 110 plane and $5·5 \times 10^{14}$ for the 100 plane. If the surface is not smooth, these numbers must be multiplied by a factor ρ which is greater than unity.

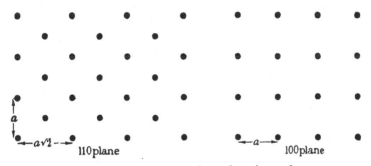

Fig. 8. Arrangement of atoms in the surface planes of tungsten.
$a = 3·15 \times 10^{-8}$ cm.

The total numbers of molecules adsorbed in five independent experiments were respectively 4·2, 4·4, 4·3, 3·3 and $3·8 \times 10^{14}$ respectively. We take the largest $4·4 \times 10^{14}$ as being nearest to the correct value and assume that the lower values are due to slight contamination of the wire before the hydrogen was admitted. We compare this with what would be expected for different types of adsorbed film on an area of 0·55 cm.[2]. We shall consider three types:

(a) If the hydrogen were adsorbed as undissociated molecules with one molecule for each tungsten atom, the total number of molecules adsorbed would be $\rho \times 7·8 \times 10^{14}$ for the 110 plane and $\rho \times 5·5 \times 10^{14}$ for the 100 plane.

(b) If the hydrogen were adsorbed as undissociated molecules, and if the occupation of a given site made it impossible for any of the four neighbouring sites to be occupied, the number of molecules adsorbed would be $\rho \times 3·9 \times 10^{14}$ for the 110 plane and $\rho \times 2·75 \times 10^{14}$ for the 100 plane.

(c) If the hydrogen were dissociated on adsorption, and if there were one atom for each tungsten atom in the surface, the number of adsorbed molecules would be $\rho \times 3 \cdot 9 \times 10^{14}$ for the 110 plane and $\rho \times 2 \cdot 75 \times 10^{14}$ for the 100 plane.

A reasonable value of ρ can be taken as somewhere between 1 and 2. Thus for either type of surface plane both (b) and (c) are in accord with the experimentally determined number of adsorbed molecules. It should be mentioned here that we shall see in §§ 4·1 and 6·3 that the numbers in (b) should be reduced by about 20 % and those in (c) by about 8 %.

It is usually assumed that hydrogen adsorbed on metals is either held by van der Waals forces in the molecular form or else dissociated and the two atoms held separately by chemical forces, and Lennard-Jones (1932) has considered this point of view in detail. The high heat of adsorption discussed in the next section and the stability of the film show that here we have to do with chemisorption, and we shall assume as a working hypothesis that it is (c) or adsorption with dissociation that occurs. The possibility of (b), i.e. the chemisorption of undissociated molecules, cannot be excluded by these experimental results, but its occurrence is unlikely. This process will be discussed in Chapter 6. One of the main objects of future work must be to decide whether (b) or (c) is the actual process, particularly for other gases.

In connexion with (c) it can be seen from fig. 8 that, for both the 110 and the 100 planes, each site for adsorption has four nearest neighbours, and that no two neighbours of a given site are neighbours of each other. The importance of this will be seen in §2·3.

1·5. The Variation of Heat of Adsorption of Hydrogen on Tungsten with the Fraction of Surface Covered

We assume that in the final film $4 \cdot 4 \times 10^{14}$ molecules are adsorbed and that variations in the amount adsorbed in different experiments are due to slight initial contamination of the wire. In fig. 9 the measured values of the heat of adsorption are plotted against the ratio of n_f, the number of filled places, to $8 \cdot 8 \times 10^{14}$ for the five different experiments, a different sign being used for each experiment. The relative values of the heats for the successive admissions in a particular experiment can be fixed with a much higher accuracy than the absolute values which are used in comparing different experiments. The results indicate a relation between heat of adsorption and amount adsorbed, which is nearly linear.

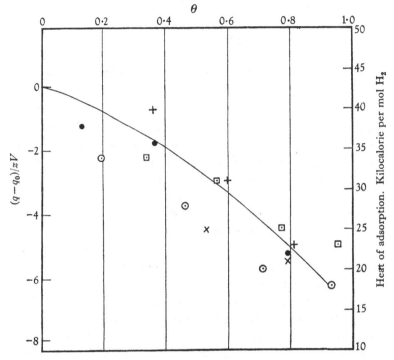

Fig. 9. Relation between heat of adsorption and amount adsorbed, taking the mean number of occupied sites in the complete film to be $8\cdot8 \times 10^{14}$. The full line is the theoretical curve calculated (§ $2\cdot6$) for an immobile film in which each particle occupies two closest neighbour sites.

A consideration of the results collected in fig. 9 indicates immediately the theoretical developments that are necessary to make a full interpretation possible. It cannot be supposed that the large and regular change in the heat of adsorption as the surface becomes covered is, for a smooth wire surface, due to variations in the activity of different parts. There must therefore be strong repulsive forces between the adsorbed particles, and to understand the phenomena involved it is necessary to develop the theory of adsorption when such repulsive forces are taken into account. We shall discuss this problem in the next chapter.

CHAPTER 2

THE THEORY OF HEAT OF ADSORPTION WITH INTERACTION BETWEEN ADSORBED PARTICLES

2·1. Langmuir's Theory of Adsorption on Definite Sites

A fundamental advance in the theory of adsorption was made when Langmuir (1912, 1915, 1926) introduced the idea that, owing to the regular arrangement of the atoms in the surface of a solid, the force acting on a gas atom near the surface would not depend only on the distance of the gas atom from the surface, but would vary over the surface so as to be much stronger at some points than at others. He proposed that the effect of this could be taken into account by assuming that there is a definite number, denoted by n_s, of sites per unit area of solid surface on which adsorption of gas atoms or molecules can take place, and that, when these sites are all occupied, the force acting on a molecule approaching the surface is so small that no further adsorption occurs. Calculations by Lennard-Jones of the potential field near a solid surface (1928, 1932) indicate that such a picture is reasonable. Its correctness has been shown more recently by Tompkins and Crawford (1948). In experiments on the adsorption of polar and non-polar gases on barium fluoride crystals it has been found that the amount of gas adsorbed in a monolayer is independent of the particular gas adsorbed. In fact, even though the cross-sectional areas of nitrous oxide and ammonia molecules differ by 50%, the same amount of each was adsorbed in a complete monolayer. Thus the number of molecules adsorbed in a monolayer depends not on the size of the adsorbed molecules, which do not form a close-packed structure on the surface, but on the sites for adsorption defined by the potential field provided by the solid surface.

We suppose that each gas molecule that is adsorbed occupies one site, and it is convenient as an introduction to our problem to consider the simplest application of this idea to the deduction of the Langmuir adsorption isotherm in which the effects of forces between adsorbed particles are neglected. Let a fraction θ of the sites be occupied when equilibrium is established with a gas at pressure p dynes cm.$^{-2}$ and temperature T° K. The number of molecules striking unit area of the solid per second is $p(2\pi mkT)^{\frac{1}{2}}$. If the condensation coefficient, that is, the fraction of those molecules striking vacant sites which condense, be denoted by α, then the number condensing on unit area per second is

$$\alpha(1 - \theta)p(2\pi mkT)^{-\frac{1}{2}},$$

where θ is the fraction of the sites that are occupied. If we assume that there is no interaction between adsorbed particles, the number evaporating per second is proportional to the number present, i.e. to θ, and is put equal to $A\theta$ per unit area, where A is a function of the temperature. For equilibrium the rates of condensation and of evaporation are equal. Thus

$$\frac{\theta}{1 - \theta} = Bp, \qquad (2\cdot1)$$

where B is a function of the temperature. This is the Langmuir adsorption isotherm.

2·2. Mobile and Immobile Adsorbed Films

When we begin to formulate the theory of the heat of adsorption, taking into account the effects of the interactions between adsorbed particles, it is at once evident that there will be a profound difference between the behaviour of mobile and immobile films, and it will be shown that the results plotted in fig. 9 are inconsistent with what would be expected of a mobile film but are in accord with what would be expected of an immobile one.*

By a mobile film is meant one in which the energy of activation necessary to enable an adsorbed particle to move from a given site to a neighbouring vacant site is much less than kT, so that the particles move freely from one site to another. This free movement enables the film to take up equilibrium configurations

---

* Some authors appear to attempt to contrast 'localized' with 'mobile'. Since, however, for an array of localized sites, adsorbed films of the two types specified here can be distinguished, it is preferable to use the word 'localized' to describe the array of sites and the contrasting words 'mobile' and 'immobile' to describe the types of film that can be adsorbed on an array of localized sites. Any other usage appears to be unnecessarily confusing (see Miller 1947).

during the occurrence of any process. Thus for a mobile film we assume that at each instant the particles on the surface may be treated as having an equilibrium Boltzmann distribution under the influence of the forces between them.

For an immobile film, on the other hand, we suppose that the energy of activation necessary to enable a particle to move from one site to another is so much greater than kT that, for the times with which we are concerned in any given experimental investigation, the particles may be treated as remaining fixed on the sites on which they are first adsorbed. For such a film the distribution at any instant will not necessarily be an equilibrium distribution.*

These two extreme cases can be dealt with theoretically. The methods of statistical mechanics can be applied to the equilibrium distribution in the mobile film and to some problems for the immobile film. Intermediate cases undoubtedly occur in which the particles can move from site to site but in which there is an appreciable time lag in setting up the equilibrium distribution if this is disturbed in any way.

The following simple physical considerations illustrate how the behaviour of mobile and immobile films may be different. Consider the simplest case of all, the formation of a film in which there is repulsive interaction between adsorbed particles and in which each adsorbed molecule occupies one site on the surface. We suppose that the interaction is appreciable only between particles adsorbed on neighbouring sites, and that adsorption occurs on the simple quadratic lattice of sites formed by the intersections of the lines in fig. 10. The adsorbed particles are represented by the black dots, and it will be seen that, with the arrangement represented, half the sites are occupied and no two neighbouring sites are occupied, i.e. $\theta = 0.5$ and the interaction energy is zero.

This is the state of lowest energy for $\theta = 0.5$ and, if the film is mobile, the mutual repulsive forces between the adsorbed particles will tend to make them arrange themselves in this state of

---

* It should be mentioned here that, apart from mobility of the adsorbed particles, evaporation and condensation will set up an equilibrium distribution of particles on the surface. In general, however, the temperature at which free mobility sets in will be lower than that at which appreciable evaporation occurs, so that the assumption that mobility is negligible implies the assumption that evaporation is also negligible.

lowest energy, but in accordance with Boltzmann's law the thermal motion of the particles will cause a configuration of higher energy to be taken up. If we neglect the effect of the thermal motion and assume that the system is always in the state of lowest energy, the heat of adsorption will remain constant at, say, q_0 calories per particle adsorbed up to $\theta = 0.5$. It will then fall and remain constant at $q_0 - 4V$ from $\theta = 0.5$ to $\theta = 1$, where V is the interaction energy between two particles adsorbed on neighbouring sites. If the

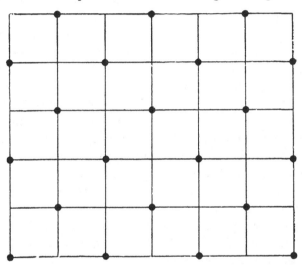

Fig. 10. State of lowest energy for adsorbed film on quadratic lattice with repulsion between particles adsorbed on neighbouring sites.

heat is plotted against θ the curve will have a sharp step at $\theta = 0.5$ (see curve (a), fig. 11). The only effect of the Boltzmann distribution will be to round off the sharp corners as shown in the other curves in fig. 11.

For the immobile film, on the other hand, if we assume that the probability that a molecule condenses when it strikes a vacant site is independent of the state of occupation of neighbouring sites, the distribution of the particles on the surface will be random, and the average number of occupied sites around any vacant site on which adsorption takes place will be strictly proportional to θ. Thus the heat of adsorption will be accurately a linear function of θ, changing from q_0 at $\theta = 0$ to $q_0 - 4V$ at $\theta = 1$. Other illustrations will be given later.

2·3. The Equilibrium Distribution of Particles on the Surface in a Mobile Film

In order to determine how the heat of adsorption varies as more and more of the sites on the surface become occupied, it is necessary to know how the energy of the adsorbed film varies with θ. That is, if the interaction energy is appreciable only between particles adsorbed on neighbouring sites, it is necessary to know, for a given value of θ, the average number of pairs of interacting particles on the surface.. The method of solving such problems, which we shall now discuss, is based on that first given by Peierls (1936) in connexion with a development of some work by Fowler (1936) on critical condensation conditions when there are attractive forces between adsorbed particles.

We shall suppose that there is one layer of N_s possible sites for adsorption arranged in a regular array in such a way that each site has z nearest neighbours, and that there is an interaction energy V for each pair of neighbouring adsorbed atoms, V being positive if the force between the particles is repulsive. For particles not adsorbed on neighbouring sites we assume that the interaction energy is negligible. This assumption is justified for all forces except dipole forces, which are discussed in Chapter 7.

Let there be θN_s occupied sites in all, and let us consider various arrangements of these occupied sites. If, in any given arrangement, X pairs of neighbouring sites are occupied, the relative probability of such an arrangement is, by Boltzmann's law,*

$$e^{-XV/kT} = \eta^{X}, \qquad (2\cdot2)$$

where η is defined by $\qquad \eta = e^{-V/kT}. \qquad (2\cdot3)$

We select a group of sites consisting of site o, which we call the central site, and sites 1 to z, its z neighbours, which we call the first shell, and we define a set of numbers $\theta_0, \theta_1, \dots, \theta_z$ which are 1 or o according to whether the corresponding sites are occupied or unoccupied.

---

* This is true only if we assume that the vibrational partition function $v_s(T)$ for an adsorbed particle is not affected by the state of occupation of neighbouring sites. We make this assumption throughout. A full examination of its validity is important, and a first step in this direction has been made by Rushbrooke (1938) for the case in which there are attractive forces between adsorbed particles.

The problem is to determine the relative probability that, for a given value of θ, the parameters $\theta_0, \theta_1, \ldots, \theta_z$ shall have an assigned set of values. Consider the $N_s - z - 1$ outer sites separately. When $\theta_0 + \theta_1 + \ldots + \theta_z$ changes, the number of particles on the outer sites is also altered, and this affects both the interaction energy of the particles on the outer sites and the number of ways of arranging them. When all the $\theta N_s = M$ particles are adsorbed on the outer sites let the average total interaction energy be E_0. Further, let the average interaction energy per particle adsorbed on first shell sites between them and the outer sites be $V_1(\theta)$. For the assigned set of values of $\theta_0, \theta_1, \ldots, \theta_z$ let there be x pairs of closest neighbour sites occupied amongst the selected group. Then the average interaction energy for a fixed value of θ and the assigned set of values of $\theta_0, \theta_1, \ldots, \theta_z$ is

$$xV + (\theta_1 + \ldots + \theta_z) V_1(\theta) + E_0 - (\theta_0 + \theta_1 + \ldots + \theta_z)\frac{dE_0}{dM}. \quad (2\cdot4)$$

For a given value of θ the relative probability of this assigned set of values of $\theta_0, \theta_1, \ldots, \theta_z$ is

$$\eta^x \zeta^{\theta_1 + \ldots + \theta_z} \xi^{\theta_0 + \theta_1 + \ldots + \theta_z}. \quad (2\cdot5)$$

This can be written $\eta^x \xi^{\theta_0} \epsilon^{\theta_1 + \ldots + \theta_z},$ (2·6)

where ϵ has been written for $\xi\zeta$ and

$$\zeta = \exp\{-V_1(\theta)/kT\}, \quad \xi = \exp\left\{\frac{dE_0}{dM}\bigg/kT\right\}. \quad (2\cdot7)$$

In effect, this method of attack is equivalent to the approximation used by Bethe (1935) in the theory of order–disorder phenomena in alloys. Some discussion of the nature of the approximation and an indication that expression (2·6) does give the correct form for the relative probability can be found in an analysis of the Bethe approximation which has been given by Guggenheim (1938). By writing down all the possible combinations of values of $\theta_0, \theta_1, \ldots, \theta_z$ and using (2·6) to give the relative probability of each, we can obtain an expression in terms of ϵ for the probability of finding one of the first shell sites, say site 1, occupied. Since every site must be an average site this average value must be equal to θ. This condition fixes the value of ϵ.

We shall carry out the detailed calculations for the case in which no two neighbours of a given site are neighbours of each other. Practically, this restricts the application of the results to $z=2$, i.e. to linear chains which are of trivial importance, and to $z=4$, which, as has been pointed out in § 1·4, is the application we require. The importance of this restriction is that with it

$$x = \theta_0(\theta_1 + \ldots + \theta_z), \qquad (2·8)$$

and we obtain simple algebraical expressions for all our quantities instead of having to tabulate the results and use numerical methods. Using (2·8) in (2·6), the relative probability of a given set of values of $\theta_0, \ldots, \theta_z$ becomes

$$\xi^{\theta_0} \eta^{\theta_0(\theta_1 + \cdots + \theta_z)} \epsilon^{\theta_1 + \cdots + \theta_z}. \qquad (2·9)$$

We now determine ϵ by calculating the average occupation of site 1. If $\theta_0 = 0$, the average value of θ_1 is given by

$$\frac{\sum \epsilon^{1 + \theta_2 + \cdots + \theta_z}}{\sum \epsilon^{1 + \theta_2 + \cdots + \theta_z} + \sum \epsilon^{\theta_2 + \cdots + \theta_z}},$$

the sums* being for all combinations of values of $\theta_2, \ldots, \theta_z$; that is, the average value of θ_1 is given by

$$\epsilon/(1 + \epsilon).$$

Similarly, if $\theta_0 = 1$, the average value of θ_1 is

$$\eta\epsilon/(1 + \eta\epsilon).$$

Taking the two results together, the overall average value of θ_1 is

$$(1 - \theta)\frac{\epsilon}{1 + \epsilon} + \theta\frac{\eta\epsilon}{1 + \eta\epsilon}.$$

This must be equal to θ and, therefore,

$$\frac{\theta}{1 - \theta} = \frac{\epsilon(1 + \eta\epsilon)}{1 + \epsilon}. \qquad (2·10)$$

For a given value of η (i.e. of V/kT) this equation enables us to obtain ϵ as a function of θ and gives us all the results we require for the calculation of heats of adsorption.

Let us consider some examples of the application of these results. If $\eta = 4 \times 10^{-2}$, which (see § 2·6) corresponds approximately to $e^{-V/kT}$ at 700° K. for a film of hydrogen if it is atomic, we obtain from equation (2·10) that, at $\theta = 0·4$, $\epsilon = 1·67$, and, at $\theta = 0·6$, $\epsilon = 15$. If at $\theta = 0·4$ the central site is occupied, the probability that site 1 is occupied is $\eta\epsilon/(1 + \eta\epsilon)$ or 0·062; if the system

* $\sum \epsilon^{\theta_1 + \theta_2 + \cdots + \theta_z} = {}_zC_0 + {}_zC_1\epsilon + {}_zC_2\epsilon^2 + \ldots + {}_zC_z\epsilon^z = (1 + \epsilon)^z.$

were in the state of lowest energy discussed in § 2·2, this probability would be zero, while for a random distribution (i.e. no interaction between adsorbed particles) it would be 0·4. If at $\theta = 0·6$ the central site is vacant, the probability that site 1 is vacant is $1/(1+\epsilon)$ or 0·063; if the system were in the state of lowest energy, this probability would be zero, while for a random distribution it would be 0·4.

Although we do not need it at present, it is convenient to consider here the determination of ξ. To do this we calculate the average value of θ_0. This is

$$\frac{\Sigma \xi(\eta\epsilon)^{\theta_1 + \cdots + \theta_z}}{\Sigma \xi(\eta\epsilon)^{\theta_1 + \cdots + \theta_z} + \Sigma \epsilon^{\theta_1 + \cdots + \theta_z}},$$

the sums as before being taken for all possible combinations of values of $\theta_1, \ldots, \theta_z$. The average value of θ_0 must also be equal to θ, and we have therefore

$$\frac{\xi(1+\eta\epsilon)^z}{\xi(1+\eta\epsilon)^z + (1+\epsilon)^z} = \theta,$$

or

$$\frac{\theta}{1-\theta} = \xi\left(\frac{1+\eta\epsilon}{1+\epsilon}\right)^z. \qquad (2·11)$$

Dividing this by (2·10) we obtain

$$\xi = \epsilon\left(\frac{1+\epsilon}{1+\eta\epsilon}\right)^{z-1}. \qquad (2·12)$$

After ϵ has been determined as a function of θ from (2·10), equation (2·12) enables us to determine ξ as a function of θ.

2·4. The Variation of Heat of Adsorption with Fraction of Sites occupied for Simple Adsorption into a Mobile Film

The application of these results in connexion with heats of adsorption has been given by Wang (1937). The simplest is to a system in which each adsorbed particle occupies one site and there is no dissociation. The heat of adsorption per molecule q is the decrease in the energy of the system when one molecule is adsorbed, and is given by

$$q = u - \frac{\partial U}{\partial (N_s \theta)}, \qquad (2·13)$$

where u is the energy of a molecule in the gas phase and U the total energy of the molecules in the adsorbed film. $N_s\theta$ is, of course, the total number of adsorbed molecules. U is given by

$$U = U_0 + \overline{X}V, \qquad (2\cdot14)$$

where U_0 is the energy of the adsorbed molecules apart from the effect of interactions and $\overline{X}$ is the average number of interacting pairs of molecules for a given value of θ. $\overline{X}$ is equal to $\frac{1}{2}zN_s\theta$

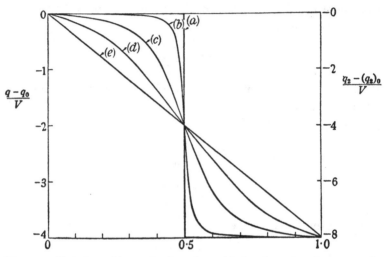

Fig. 11. Variation of heat of adsorption with fraction of surface covered. Curve (a) $\eta = \mathrm{e}^{-V/kT} = 0$; (b) $\eta = 3\cdot6 \times 10^{-3}$; (c) $\eta = 8\cdot2 \times 10^{-2}$; (d) $\eta = 0\cdot368$; (e) $\eta = 1$. The scale on the left is for the heat of adsorption q when there is no dissociation, and that on the right is the heat of adsorption for the case in which a diatomic molecule dissociates on adsorption. The subscript 0 refers to $\theta = 0$, and V is the interaction energy between particles adsorbed on neighbouring sites. The calculations are for $z = 4$.

times the average value of θ_1 when $\theta_0 = 1$, the factor one-half being introduced so that each pair shall be reckoned once only. That is

$$\overline{X} = \frac{1}{2}zN_s\theta\{\eta\epsilon/(1 + \eta\epsilon)\}.$$

Using (2·10) this becomes

$$\overline{X} = \frac{1}{2}zN_s\left[\theta - \frac{1 - \{1 - 4(1 - \eta)\theta(1 - \theta)\}^{\frac{1}{2}}}{2(1 - \eta)}\right]. \qquad (2\cdot15)$$

Using (2·15) in (2·13) and (2·14), we obtain

$$\frac{q - q_0}{zV} = -\frac{1}{2}\left[1 - \frac{1 - 2\theta}{\{1 - 4(1 - \eta)\theta(1 - \theta)\}^{\frac{1}{2}}}\right], \qquad (2\cdot16)$$

where q_0 is the heat of adsorption of a molecule on a bare surface, i.e. it is the heat of adsorption when $\theta = 0$.

Values of $(q - q_0)/V$ for $z = 4$ and for various values of η are plotted against θ in fig. 11. The curve for $\eta = 0$ corresponds to the case discussed in § 2·2 when the system is always in the state of lowest energy, and that for $\eta = 1$ corresponds to a random distribution.

2·5. Variation of Heat of Adsorption with Fraction of Sites occupied in a Mobile Film for Adsorption with Dissociation

The results in the last paragraph enable us to obtain the variation of the heat of adsorption q_2 of a diatomic molecule which dissociates on adsorption. This is defined as the decrease in energy when one molecule is adsorbed. If q_1 is the heat of adsorption of an atom,
$$q_2 = 2q_1 - q_d, \tag{2·17}$$
where q_d is the heat required to dissociate a molecule. From equation (2·16)
$$\frac{q_1 - (q_1)_0}{zV} = -\frac{1}{2}\left[1 - \frac{1 - 2\theta}{\{1 - 4(1-\eta)\,\theta(1-\theta)\}^{\frac{1}{2}}} \right],$$
where $(q_1)_0$ is the heat of adsorption of an atom on the bare surface and V is the interaction energy between two *atoms* adsorbed on neighbouring sites. Using this in equation (2·17) we obtain
$$\frac{q_2 - (q_2)_0}{zV} = -\left[1 - \frac{1 - 2\theta}{\{1 - 4(1-\eta)\,\theta(1-\theta)\}^{\frac{1}{2}}} \right], \tag{2·18}$$
where $(q_2)_0$ is the heat of adsorption of a molecule on the bare surface. Apart from the factor one-half this is the same as equation (2·16), and the curves plotted in fig. 11 for $z = 4$ apply to this case also, provided the scale of ordinates on the right-hand side of the figure is used.

In comparing these calculations with the experimental results for hydrogen plotted in fig. 9, we note that $(q_2)_1 - (q_2)_0 = 8V$. The difference between the heats of adsorption at $\theta = 0$ and $\theta = 1$ is 27,000 calories per mol. This gives $V = 2\cdot3 \times 10^{-13}$ erg per pair of atoms,* and at $300°$ K., the temperature at which the experi-

---

* The justification for assuming that in this case only nearest neighbours interact is as follows. The only long-range forces that are likely to arise are

ments were carried out, $\eta = e^{-V/kT} = 3 \cdot 6 \times 10^{-3}$. Thus, if the film were mobile, we should expect a relation between heat of adsorption and θ like curve (b) in fig. 11. A comparison with the experimental results in fig. 9 shows immediately that the actual variation is nothing like this, and we therefore examine the behaviour that would be expected if the film were not mobile but immobile.

2·6. The Theory of Heat of Adsorption for an Immobile Film

For an immobile film, in which each particle occupies one site and in which the probability of condensation on a vacant site is independent of the state of occupation of neighbouring sites, we have at any stage a geometrically random distribution of occupied sites, and we have already seen in § 2·2 that the heat of adsorption is a linear function of θ. If we are considering a process in which diatomic gas molecules dissociate on adsorption and the two atoms occupy two neighbouring vacant sites on the surface, we do not have a random distribution of occupied sites, because, if a given site is occupied, we know that one of the neighbouring sites must be occupied by the other atom of the diatomic molecule. In this case we have a geometrically random distribution of *pairs* of occupied sites. We now consider the relevant properties of such a distribution.

The random distribution of pairs of occupied sites was originally discussed by Roberts (1938 *a*) using an empirical method. A model of the surface containing one hundred numbered sites was used, and pairs of closest neighbour sites were chosen at random. Taking precautions to eliminate edge effects it was possible to obtain an empirical expression for the heat of adsorption. Subsequently, statistical methods were applied by Roberts and Miller (1939) to some of the properties of immobile films of mole-

forces between electrostatic dipoles. Assuming that the whole variation in the heat was due to repulsion between dipoles Roberts (1937) has shown that the dipoles would be of such strength that the contact potential between bare tungsten and hydrogen-covered tungsten would be 5·2 V. Bosworth (1937) has measured this and shown that it is 1·04 V. Thus the dipoles are about one-fifth as strong as would be required to account for the whole thermal effect and, since the interaction energy is proportional to the square of the dipole strength, electrostatic forces would account for only about 4% of the whole. The remaining 96% is due to short-range forces which can be assumed to be effective only between nearest neighbours.

cules each of which occupied two sites on the array, and the fundamental statistical relations derived by them have been used (Miller 1947) to obtain an expression for the heat of adsorption. It is no more difficult to consider the general case, and this we now do.

The Bethe method was used considering a group of sites which consisted of a central site and its four closest neighbours, the first shell sites. Let n_0, which may be 0 or 1, specify the state of occupation of the central site and n_1 be the number of occupied first shell sites apart from the site which is occupied by a molecule on the central site, if there is one. The various possible distributions for a square array are represented in table 1, in which $g(n_0, n_1)$ is the weight of the distribution specified by n_0 and n_1, i.e. it is the number of independent ways in which the distribution can occur.

TABLE 1

n_0	1	1	1	1	0	0	0	0	0
n_1	0	1	2	3	0	1	2	3	4
$g(n_0, n_1)$	4	12	12	4	1	4	6	4	1

If two neighbouring sites are occupied by atoms of different molecules it is assumed that there is an interaction energy V which is positive if it arises from repulsion. For a given value of θ, the fraction of the total number of sites that is occupied, the relative probability of any given arrangement is, on the basis of the Bethe approximation,

$$g(n_0, n_1)\, \epsilon_0^{n_0} \epsilon_1^{n_1} \eta^{n_0 n_1}, \tag{2.19}$$

where η has been written for $e^{-V/kT}$, and ϵ_0 and ϵ_1 include factors which take account of interactions between particles on the selected group of sites and particles on outer sites. The grand partition function can then be written as

$$\Xi = \sum_{n_0, n_1} g(n_0, n_1)\, \epsilon_0^{n_0} \epsilon_1^{n_1} \eta^{n_0 n_1},$$

which, using the values in table 1, reduces to

$$\Xi = \epsilon_0 (1 + \eta \epsilon_1)^{z-1} + (1 + \epsilon_1)^z. \tag{2.20}$$

The condition that the central site should be an average site is expressed by the relation,

$$\theta = \epsilon_0 \frac{\partial \log \Xi}{\delta \epsilon_0},$$

which leads to the equation

$$\frac{\theta}{1-\theta} = \frac{\epsilon_0(1+\eta\epsilon_1)^{z-1}}{(1+\epsilon_1)^z}. \tag{2.21}$$

The probability that the central site is occupied is θ, and, if it is occupied, the average number of particles on first shell sites is

$$\epsilon_1\left(1 + \frac{\partial \log \Xi_1}{\partial \epsilon_1}\right) = \frac{1+z\eta\epsilon_1}{1+\eta\epsilon_1}, \tag{2.22}$$

where Ξ_1 has been written for the first term of equation (2.20). The probability that the central site is vacant is $1-\theta$, and, if it is vacant, then the average number of particles on the first shell sites is

$$\epsilon_1 \frac{\partial \log \Xi_2}{\partial \epsilon_1} = \frac{z\epsilon_1}{1+\epsilon_1}, \tag{2.23}$$

where Ξ_2 has been written for the second term in equation (2.20). Since the first shell sites are necessarily average sites, the overall average number of particles on the first shell sites must be $z\theta$. After rearrangement of the terms, equations (2.22) and (2.23) yield

$$\frac{\theta}{1-\theta} = \frac{z}{z-1} \frac{\epsilon_1(1+\eta\epsilon_1)}{1+\epsilon_1}. \tag{2.24a}$$

The interaction energy of the adsorbed film is given by

$$U_i = V\overline{X}, \tag{2.25}$$

where $\overline{X}$ is the average number of pairs of closest neighbour sites which are occupied by different molecules. To determine the value of $\overline{X}$ it is necessary to answer the question: If the central site is occupied, what is the probability that a given first shell site will be occupied by some *other* molecule? It is clear that we require the average value of n_1 when n_0 is unity, say $(\overline{n}_1)_{n_0=1}$. From equation (2.22) this is given by

$$(\overline{n}_1)_{n_0=1} = \frac{(z-1)\eta\epsilon_1}{1+\eta\epsilon_1}, \tag{2.26a}$$

since it is one less than the average number of particles on first shell sites. The interaction energy of the monolayer is therefore given by

$$U_i = \tfrac{1}{2}N_s V\theta(z-1)\frac{\eta\epsilon_1}{1+\eta\epsilon_1}, \tag{2.27a}$$

where the factor one-half is introduced so that no interaction is counted more than once. The heat of adsorption per molecule is defined as the decrease in the energy of the whole system when one particle is transferred from the gas phase to the adsorbed phase. If q is the heat of adsorption per molecule and q_0 is the heat of adsorption of a solitary gas atom on a bare surface, then

$$q - q_0 = -\frac{\partial U_i}{\partial N_a},$$

where N_a is the number of adsorbed molecules and is equal to $\frac{1}{2}N_s\theta$. Thus

$$\frac{q - q_0}{zV} = -\frac{z-1}{z}\frac{\eta\epsilon_1}{1+\eta\epsilon_1} - \frac{z-1}{z}\theta\frac{\eta}{(1+\eta\epsilon_1)^2}\frac{\partial\epsilon_1}{\partial\theta}. \qquad (2\cdot28)$$

Differentiating equation $(2\cdot24a)$ with respect to θ we obtain

$$\frac{\partial\epsilon_1}{\partial\theta} = \frac{z-1}{z}\frac{(1+\epsilon_1)^2}{1+2\eta\epsilon_1+\eta\epsilon_1^2}\frac{1}{(1-\theta)^2}. \qquad (2\cdot29)$$

In the case of an immobile film there is a random distribution of particles. This corresponds to the case $\eta = 1$. Thus for an immobile film the average number of other molecules which occupy first shell sites when the central site is occupied is given by

$$(\overline{n_1'})_{n_0=1} = (z-1)\frac{\epsilon_1}{1+\epsilon_1}, \qquad (2\cdot26b)$$

with, in this case,
$$\epsilon_1 = \frac{z-1}{z}\frac{\theta}{1-\theta}. \qquad (2\cdot24b)$$

Thus, the interaction energy of an immobile film is given by

$$U_i = \frac{1}{2}N_s\theta(z-1)\frac{\epsilon_1}{1+\epsilon_1}V = \frac{1}{2}N_sV\frac{(z-1)^2\theta^2}{z-\theta}. \qquad (2\cdot27b)$$

It follows that the heat of adsorption of an immobile film of particles each of which occupies two sites is given by

$$\frac{q-q_0}{zV} = -\frac{(z-1)^2}{z}\frac{\theta(2z-\theta)}{(z-\theta)^2}. \qquad (2\cdot30)$$

This curve is plotted in fig. 12; the curve is not continued beyond $\theta = 0\cdot9$ for the following reason. It has been shown by Roberts and Miller (1939) that for an immobile film of this kind no further adsorption can take place into the first layer after about 92 % of the sites are occupied (see § 4·1). As the film is formed, some sites which are surrounded by four occupied sites are left vacant; for an immobile film in which each particle occupies two sites, these vacant sites can never be occupied in the first layer. Owing to this cause, adsorption into the first layer stops when there are still about 8% of the total number of sites vacant, each of these vacant sites being completely surrounded by occupied sites. It will be noticed that the curve given by equation (2·30) departs only slightly from a

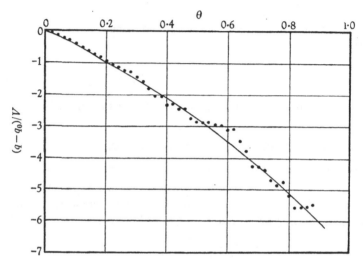

Fig. 12. Comparison of the theoretical values of the heat of adsorption of an immobile film in which each particle occupies two closest neighbour sites calculated by Miller (1947) from equation (2·30) (full curve), with the points calculated by Roberts (1938 a) using an empirical method (represented by dots).

straight line, which is the form of the variation of the heat of adsorption for an immobile film of particles each of which occupies only one site.

As has already been pointed out, this curve was originally obtained by an empirical method; in fig. 12 the points which were calculated in this way are plotted on the same diagram as the curve which is calculated from equation (2·30). It will be noticed that the empirical method used by Roberts gave quite a good approximation to the heat curve.

This heat curve can be compared with the experimental results which were obtained by Roberts. The heat curve calculated from equation (2·30) is shown as the full line in fig. 9, in which the scale in kilocalories per mol. for the experimental results is shown on the right-hand side and that for the curve calculated from equation (2·30) on the left-hand side. It will be seen that, considering the spread of the experimental results, there is reasonably good agreement. The alternative theoretical curve is that corresponding to a mobile film and shown as curve (b) in fig. 11; as has been noted at the end of § 2·5, it is evident that such a curve departs greatly from the experimental results, and it must be concluded from this theory of the heat of adsorption with a fixed interaction energy between neighbouring adsorbed particles that the hydrogen film adsorbed on tungsten at room temperatures is immobile.

2·7. It has been seen that to account for the total variation of the heat of adsorption of a hydrogen film on tungsten on the assumption that a mobile film is formed would imply a value of the interaction energy corresponding to curve (b) of fig. 11. This curve is totally different from the very nearly linear variation of the heat of adsorption which is found for an immobile film of the atoms of a diatomic molecule and shown by the full line in fig. 9. For an immobile film of particles, each of which occupies only one site on adsorption, the variation of the heat of adsorption with the fraction of the surface covered is exactly linear. The difference between these two cases should perhaps be emphasized, as failure to appreciate it* can lead to wrong conclusions.

This behaviour is different from what is found for a liquid mixture. Assuming a random distribution of molecules in a binary mixture, the number, $\overline{X}$, of closest neighbour interactions of the two kinds of molecules is given, in terms of the numbers of molecules of each kind, by

$$(N_a - \overline{X})(N_b - \overline{X}) = \overline{X}^2. \qquad (2·31)$$

A better approximation to an actual liquid mixture can be obtained by using the ideas of chemical kinetics in the form of the condition of quasi-chemical equilibrium and to write

$$(N_a - \overline{X})(N_b - \overline{X}) = \overline{X}^2 e^{-2w/kT}, \qquad (2·32)$$

where w is the energy of mixing. The vapour pressure equations which are obtained from these two formulae do not differ very

* See Miller (1948a).

greatly for solutions of non-electrolytes. A similar state of affairs holds for solution of high polymers in which it is found that to take account of the energy of mixing makes little difference to the vapour pressure curves (Miller 1948 b). Thus, in examining liquid mixtures, it is legitimate to consider the case of a random mixture, given by equation (2·31), as an approximation to the case in which a non-zero energy of mixing is allowed for, equation (2·32).

Although the contrary opinion has been expressed (Fowler and Guggenheim 1939), this is not so for the two-dimensional problem of an adsorbed monolayer. In this physical problem, the immobile film (as defined in § 2·1), giving a random distribution, corresponds to equation (2·31) in the theory of liquid mixtures and the mobile film corresponds to equation (2·32). As is evident from the heat curves which have been determined in this chapter, by no stretch of the imagination could either of these be regarded as an approximation to the other. As was realized by Roberts (1935 a, 1937, 1938 a), these two treatments correspond to distinct physical cases in the adsorption problem. Recently (Halsey and Taylor 1947), in discussion of some results of Frankenburg (1944) on the adsorption of hydrogen on tungsten powders, it has been argued that, since the heat of adsorption does not show a linear variation with the fraction of the surface covered, the adsorption cannot be taking place on sites for each of which the adsorption energy is the same. This conclusion is ill-founded and is evidently due to the assumption that the case of a random distribution is a reasonable approximation to that of an equilibrium Boltzmann distribution. In fact, as far as the theory of adsorption has so far been developed, all that could be inferred from a non-linear variation of the heat of adsorption is that the adsorbed monolayer could not exist as an immobile film, but could exist as a mobile film occupying localized sites (Miller 1948 a) with a fixed interaction energy between the adsorbed particles. While still postulating that the adsorption of a single particle on any site of the surface involves the same energy of adsorption, one can allow for the facts that the Langmuir sites are the minima of a continuously varying potential field and that the interaction energy between two particles varies with their distance apart. When this calculation is carried out (Chapter 3) it is found that a non-linear variation of the heat of adsorption with the fraction of the surface covered, does not provide a very firm basis *even* for rejecting the possibility of an immobile adsorbed film.

The point which we want to emphasize here is that, from the point of view of carrying out an experiment, one is concerned with two distinct physical cases and these give quite different heat curves. In any discussion of adsorption measurements both these possibilities must be considered separately, since as far as the variation of the heat of adsorption is concerned, the case of a random distribution of adsorbed particles cannot possibly be considered to be an approximation to the case of a Boltzmann distribution of adsorbed particles.

CHAPTER 3

VARIATION OF THE POTENTIAL ENERGY
OVER THE SURFACE

3·1. Introduction

So far it has been assumed that there is a definite number of sites per unit area of the surface on which adsorption can take place, that between two particles adsorbed on neighbouring sites there is a fixed interaction energy, and that between particles at a greater distance apart this interaction energy is negligible. For an actual surface over which the potential energy varies continuously and on which the sites for adsorption are the positions of minimum potential energy, the assumption of a fixed interaction energy between particles adsorbed on neighbouring sites may be far from the truth when account is taken of the fact that the interaction energy of two particles is a function of the distance between them and that this distance varies with the state of occupation of neighbouring sites. The accuracy of the results given by the 'fixed interaction' method depends on the relation between the variation of the potential energy of the adsorbed atoms with their distance apart, and the variation of the potential energy of a single adsorbed particle with its distance from a site for adsorption. The greater the latter is compared with the former the more nearly correct are the results given by the fixed interaction method.

In fig. 13 the lower full curves show a two-dimensional cross-section of how the energy of a single adsorbed particle in its lowest state might be expected to vary with its position on the surface. A and B are 'sites' for adsorption. The distance between A and B is three of the arbitrary units of length used. The upper curves show the mutual potential energy of two adsorbed particles at various distances apart. The actual value V of this energy when the distance is equal to AB is the same in (a), (b) and (c), but the variation with distance is different in the three cases. In (a) it has an intermediate value, in (b) it is rapid, and in (c) it is slow. The lower dotted curves show the potential energy of two particles adsorbed in their lowest states on neighbouring sites and arranged in each case symmetrically about the point O.

If as in (b) the rate of variation with distance apart of the mutual potential energy of two particles on neighbouring sites is large compared with the rate of variation of the potential energy of a single particle with its position, the interaction energy between two

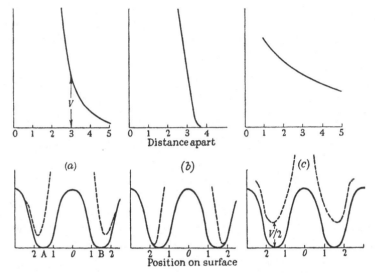

Fig. 13. The upper curves show the variation with distance apart of the potential energy of two adsorbed particles due to their mutual repulsive forces. The lower full curves show the variation in the potential energy of a solitary adsorbed particle in its lowest state at different places on the surface. The dotted curves show how this is modified when two particles in their lowest states are adsorbed on two neighbouring sites in positions symmetrical about O. If the particles are at a distance apart equal to the distance between the minima of the full curves, the interaction energy V is the same in the three cases, but the slopes of the upper curves are very different. In (a) the heat of adsorption of the second particle is about $q_0 - \frac{1}{4}V$; in (b) it is about $q_0 - \frac{1}{2}V$; in (c) it is about $q_0 - V$. The length and energy scales respectively are the same in the upper and lower figures.

particles adsorbed on neighbouring sites will be considerably less than V provided the relevant outer neighbouring sites, which are not shown on the diagram, are vacant. It will be seen that in an extreme case of this type the variation of the heat of adsorption with θ for mobile and immobile films would be more similar than in the model considered before, since for low values of θ the heat would in both cases vary slowly with θ. One step in obtaining a general idea of the behaviour was taken by Wilkins (1938) who considered the limiting case in which the potential energy of a single particle is assumed to be uniform over the surface.

If, on the other hand, the rate of variation with distance of the mutual potential energy of two particles is small, as in (c), the model used in the preceding chapter represents the actual behaviour closely. This question has been discussed, for a general variation with distance of the mutual potential energy, by Miller and Roberts (1941). Their treatment will now be considered.

3·2. Physical Model

We assume that the potential energy of a single adsorbed particle is given by
$$-\chi + X\phi(x), \tag{3·1}$$
where x is its distance from a site. $\phi(x)$ is a periodic function of period a, where a is the distance between neighbouring sites. The simplest function to choose for $\phi(x)$ is
$$\phi(x) = 1 - \cos(2\pi x/a). \tag{3·2}$$

We assume that, if two particles are adsorbed on two neighbouring sites on the surface and are a distance y apart, the potential energy arising from the forces between them is given by an inverse power law
$$Yy^{-m}, \tag{3·3}$$
and that the forces between particles adsorbed on sites a greater distance apart are negligible. When the particles are separated by a distance a this potential energy will be denoted by V_a. In terms of V_a, the potential energy between two particles at a distance y apart is
$$V_a(a/y)^m. \tag{3·4}$$
We write
$$X = V_a/a, \tag{3·5}$$
and the numerical calculations are carried out for $a = 1$. In this case, and with $\phi(x)$ specified by equation (3·2), the total height of the potential hills above the hollows in the field due to the surface is twice the mutual potential energy of two particles a distance a apart. In fig .14 (a) the full curve shows the variation in the potential energy of a single adsorbed particle in its lowest state as it moves about the surface, A and A' being sites for adsorption. The full

curve (i) in fig. 14(b) shows the variation with their distance apart of the mutual interaction energy of two adsorbed particles for an inverse seventh power law of interaction. When two particles are adsorbed on neighbouring sites there is an additional contribution to the potential energy of the system. If the two particles are in their lowest states and are in positions symmetrical about O and the surrounding sites are vacant, this additional part, which is due to the potential field provided by the surface, is represented by the full curve (ii) in fig. 14(b). The resultant potential energy of the

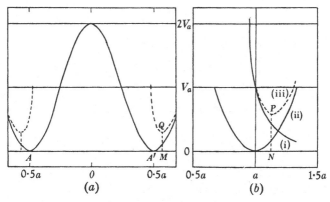

Fig. 14. The potential energy of a pair of particles in their lowest states. The full curve in (a) shows the potential energy of a single particle in its lowest state as it moves about the surface. The full curve (i) in (b) shows the variation, with their distance apart, of the mutual interaction energy of a pair of particles, for an inverse seventh power law of interaction. The full curve (ii) shows the contribution to the potential energy due to the field provided by the surface as the two particles are displaced from the sites A, A' symmetrically about O. The broken curve (iii) in (b) shows the resultant potential energy of a pair of particles, with a minimum corresponding to their equilibrium displacement. The broken curves in (a) show the resultant potential energy of each particle.

system is represented by the broken curve (iii) in fig. 14(b), and the position of the minimum of this curve gives the separation of the particles in their equilibrium positions. The resultant potential energy of each particle is represented by the broken curves in fig. 14(a). A comparison of the full and broken curves in fig. 14(a) shows how the potential field is modified by the interaction between neighbouring particles. The scales of the curves in fig. 14 (a, b) are the same. In fig. 14(a) the abscissae give the positions of the particles on the surface, and in fig. 14 (*b*) the abscissae are the distances between the two particles.

Suppose that a row of r particles occupies r sites with no gaps, and that the two sites which are the nearest neighbours of the two end sites of the row are both vacant. We have to determine the equilibrium position of each of the particles and to calculate the total energy of the particles when each is in its equilibrum position. Let the displacements of the particles from their respective sites for adsorption be $x_1, x_2, \ldots, x_r$. The potential energy of the system of particles is given by

$$-r\chi + U, \qquad (3\cdot6)$$

where the interaction energy, U, is given by

$$\frac{U}{V_a} = \sum_{s=1}^{r-1}\left(1 + \frac{x_{s+1}-x_s}{a}\right)^{-m} + \frac{1}{a}\sum_{s=1}^{r}\left(1 - \cos\frac{2\pi x_s}{a}\right). \qquad (3\cdot7)$$

For equilibrium the x_s must satisfy the equations

$$\frac{\partial U}{\partial x_s} = 0 \quad (s = 1, 2, \ldots, r). \qquad (3\cdot8)$$

When r is odd, considerations of symmetry show that the $\frac{1}{2}(r+1)$th particle is undisplaced, and that the displacement of the sth particle is minus that of the $(r-s+1)$th particle. We can then write the interaction energy in terms of the $\frac{1}{2}(r-1)$ quantities $x_1, x_2, \ldots, x_{\frac{1}{2}(r-1)}$, and we obtain

$$\frac{U}{2V_a} = \sum_{s=1}^{\frac{1}{2}(r-1)}\left\{\left(1 + \frac{x_{s+1}-x_s}{a}\right)^{-m} + \frac{1}{a}\left(1 - \cos\frac{2\pi x_s}{a}\right)\right\}, \qquad (3\cdot9)$$

where the x_s satisfy the equations

$$\left.\begin{aligned}\left(1 + \frac{x_2-x_1}{a}\right)^{-m-1} + \frac{2\pi}{am}\sin\frac{2\pi x_1}{a} &= 0, \\ -\left(1 + \frac{x_s-x_{s-1}}{a}\right)^{-m-1} + \left(1 + \frac{x_{s+1}-x_s}{a}\right)^{-m-1} + \frac{2\pi}{am}\sin\frac{2\pi x_s}{a} &= 0 \\ (s = 2, 3, \ldots, \tfrac{1}{2}(r-1)).\end{aligned}\right\} \qquad (3\cdot10)$$

When r is even, considerations of symmetry show that there is no undisplaced particle and that, as before, the displacement of the sth particle is minus that of the $(r-s+1)$th particle. We thus obtain the following expression for the interaction energy of the system:

$$\frac{U}{2V_a} = \sum_{s=1}^{\frac{1}{2}(r-2)} \left\{ \left(1 + \frac{x_{s+1} - x_s}{a}\right)^{-m} + \frac{1}{a}\left(1 - \cos\frac{2\pi x_s}{a}\right) \right\}$$
$$+ \frac{1}{2}\left(1 - \frac{2x_{\frac{1}{2}r}}{a}\right)^{-m} + \frac{1}{a}\left(1 - \cos\frac{2\pi x_{\frac{1}{2}r}}{a}\right), \quad (3\cdot11)$$

where the x_s satisfy the equations

$$\left(1 + \frac{x_2 - x_1}{a}\right)^{-m-1} + \frac{2\pi}{am}\sin\frac{2\pi x_1}{a} = 0,$$

$$-\left(1 + \frac{x_s - x_{s-1}}{a}\right)^{-m-1} + \left(1 + \frac{x_{s+1} - x_s}{a}\right)^{-m-1} + \frac{2\pi}{am}\sin\frac{2\pi x_s}{a} = 0$$

$$(s = 2, 3, \ldots, \tfrac{1}{2}(r-2)),$$

$$-\left(1 + \frac{x_{\frac{1}{2}r} - x_{\frac{1}{2}(r-2)}}{a}\right)^{-m-1} + \left(1 - \frac{2x_{\frac{1}{2}r}}{a}\right)^{-m-1} + \frac{2\pi}{am}\sin\frac{2\pi x_{\frac{1}{2}r}}{a}.$$

$$(3\cdot12)$$

Equations (3·10) and (3·12) can be solved exactly (by numerical methods), and the interaction energies of rows containing various numbers of particles can be calculated by substituting these values in equations (3·9) and (3·11). This calculation has been carried out for rows of up to nine particles for inverse seventh and inverse eleventh power laws of interaction between neighbouring particles for the case $a = 1$. The values of the interaction energies in these two cases are given in the second and third columns of table 2, and for purposes of comparison the fourth column contains the values that would be obtained for corresponding rows of particles if the interaction energy were fixed. In each case the unit is V_a ergs.

TABLE 2. *Interaction energies of rows of adsorbed particles*

No. of particles	Inverse seventh power law	Inverse eleventh power law	Fixed inter-action energy
2	0·589	0·425	1
3	1·405	1·082	2
4	2·323	1·889	3
5	3·284	2·768	4
6	4·268	3·682	5
7	5·260	4·637	6
8	6·258	5·614	7
9	7·256	6·598	8

3·3. The Variation of the Heat of Adsorption with the Fraction of the Surface Covered for a Mobile Film

We consider a row of n_s sites for adsorption. To eliminate edge effects we suppose that the two end sites are repeated at the opposite ends.* In fig. 15 the black dots represent the n_s sites for adsorption and the open circles represent the repeated sites. The various ways in which r ($r = 1, 2, \ldots, n_s$) particles can be arranged on these sites are considered. Let the interaction energy for a given arrangement, specified by the subscript i, of the particles, calculated from equations (3·9) or (3·11), be $\vartheta_i V_a$. The relative probability of this configuration is η^{ϑ_i}, where $\eta = e^{-V_a/kT}$. The

Fig. 15.   Group of r_s sites (full circles); the two end sites are repeated at the opposite ends (open circles) to eliminate end effects.

average interaction energy of the system when r particles are adsorbed on the n_s sites is given by

$$U_r = V_a \frac{\Sigma_{(i)_r} \vartheta_i \eta^{\vartheta_i}}{\Sigma_{(i)_r} \eta^{\vartheta_i}}, \qquad (3·13)$$

where the subscript $(i)_r$ is used to indicate that the summations are to be carried out over all possible configurations of a fixed number, r, of particles on the sites. This gives the interaction energy of the adsorbed particles for $\theta = r/n_s$. We take the heat of adsorption as given to a first approximation by

$$q = q_0 - \frac{1}{n_s} \frac{\delta U_r}{\delta \theta}, \qquad (3·14)$$

where q_0 is the heat of adsorption of a single particle on a bare surface. The change in interaction energy as θ changes by $\delta\theta = 1/n_s$ from r/n_s to $(r+1)/n_s$ is given by

$$\delta U_r = U_{r+1} - U_r,$$

so that $\qquad\qquad q - q_0 = U_r - U_{r+1}, \qquad (3·15)$

where q is the heat of adsorption for some value of θ intermediate between r/n_s and $(r+1)/n_s$. In graphing q against θ we assume that equation (3·15) gives the heat of adsorption at $\theta = (r + \frac{1}{2})/n_s$.

* This is equivalent to considering the sites arranged around a circle.

The calculations were carried out for inverse seventh and inverse eleventh power laws of interaction when there were in all ten and eight sites for adsorption.

The results for an inverse seventh power law of interaction are shown in curve (i) of fig. 16, and for an inverse eleventh power law

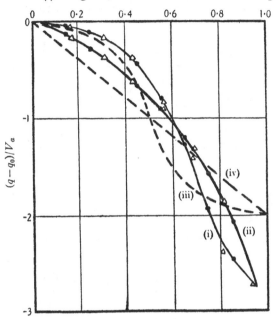

Fig. 16. Variation of the heat of adsorption with the fraction of the surface covered. Fig. 16 is for an inverse seventh power and fig. 17 for an inverse eleventh power law of interaction between neighbouring adsorbed particles. In each case, curve (i) is for a mobile film and curve (ii) for an immobile film. All the curves are drawn through the points obtained considering ten sites for adsorption (dots), and the points which are obtained when only eight sites for adsorption are considered are represented by open triangles. The broken curves (iii) and (iv) are those obtained from the simple theory assuming a fixed interaction energy between neighbouring adsorbed particles; the curves (iii) are for a mobile film and the curves (iv) are for an immobile film. Curves (i) and (iii) are calculated for $\eta = 0.1$, while curves (ii) and (iv) are independent of η.

of interaction in curve (i) of fig. 17. The points obtained when ten sites for adsorption are considered have been represented by black dots and those obtained when eight sites for adsorption are considered have been represented by open triangles. The curves are drawn through the black dots. It will be noticed that the points obtained when eight sites are considered all lie practically on the curve which is obtained when ten sites are considered; this shows that this curve is a good approximation to the actual variation. The broken curve (iii) in figs. 16 and 17 is the corresponding curve obtained when we assume that the interaction energy between particles adsorbed on neighbouring sites is fixed, the calculations being carried out for the same value of η as for

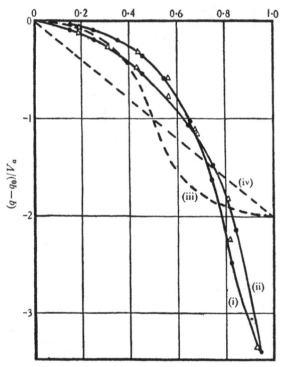

Fig. 17. See caption to Fig. 16.

the curves (i). It is important to notice that the physical model used here gives a heat curve in which the change in the heat of adsorption in the neighbourhood of $\theta = 0.5$ is much less abrupt than that obtained from the theory with fixed interaction energy, and that for a given value of V_a it gives a much greater total change in the heat of adsorption.

It will be noticed that these curves terminate at $\theta = 0.95$; this is because we have considered only a limited number of sites. It is important to investigate the behaviour of the film as $\theta \to 1$.

This has been done by determining the heat of adsorption of the last particle for various numbers of sites. The results which are obtained are given in table 3, the unit in each case being V_a ergs. Thus, the heat of adsorption of the last particle rapidly approaches a limiting value as the number of sites considered is increased; and the limits of $q_0 - q$ are about $2 \cdot 75 \ V_a$ and $3 \cdot 43 \ V_a$ respectively for inverse seventh and inverse eleventh power laws of interaction.

Further, it is clear from table 3 that the energy change as r changes from $n_s - 1$ to n_s calculated by considering a limited number of sites, is *less* than the actual change. In addition, we have assumed that equation $(3 \cdot 15)$ gives the heat of adsorption at $\theta = (r + \frac{1}{2})/n_s$, whereas all that we can assert is that equation $(3 \cdot 15)$ gives the heat of adsorption at some value of θ intermediate between r/n_s and $(r + 1)/n_s$. These two effects will slightly distort the heat curve, and the extent of the distortion can be seen from the divergence, at high values of θ, between this heat curve and the step curve calculated from the states of minimum energy* and shown in fig. 19.

TABLE 3. $(q_0 - q)/V_a$ *for the last particle for various numbers of sites*

No. of sites	Inverse seventh power law of interaction	Inverse eleventh power law of interaction
6	2·715	3·244
8	2·739	3·362
10	2·744	3·402
12	2·745	3·415
14	2·746	3·420

3·4. The Relation between the Heat of Adsorption of an Immobile Film and the Fraction of the Surface Covered by it

In the previous section it has been assumed that the particles can be treated as though they had an equilibrium Boltzmann distribution. If the activation energy necessary to enable a particle to move from one site to the next is much greater than kT, neither migration nor evaporation of particles will occur to any appreciable extent during the time of an experimental observation; and provided that in such an immobile film the probability of con-

* For a discussion of the states of minimum energy see § 3·6.

densation of a particle striking a vacant site is independent of the state of occupation of neighbouring sites, the distribution of particles on the surface at any instant is a random one.

To examine the behaviour of such a random distribution we consider all configurations of r particles on a group of n_s sites when one selected site (say site 1) is vacant, and determine the change in energy when a particle is placed on this vacant site. The average change in energy, following the occupation of site 1, is the heat of adsorption for some value of θ intermediate between r/n_s and $(r+1)/n_s$, say at $(r+\frac{1}{2})/n_s$. The change in energy for all configurations can be calculated immediately from the values of the energy of rows of adsorbed particles given in table 2. The results are plotted in the full curves (ii) in figs. 16 and 17 which refer respectively to inverse seventh and inverse eleventh power laws of interaction between neighbouring adsorbed particles. The black dots and open triangles on these curves have the same significance as the corresponding symbols on the curves for mobile films. The broken curve (iv) is the heat curve which is obtained when it is assumed that the interaction energy between particles adsorbed on neighbouring sites is fixed. It will be seen that for a given law of interaction the difference between the behaviour of a mobile and of an immobile film is much less marked for the physical model used in these calculations (curves (i) and (ii)) than it is for the physical model in which a fixed interaction energy between particles adsorbed on neighbouring sites is assumed (curves (iii) and (iv)).

3·5. Application to Experiment

These results have two important consequences in the discussion of experimental heats of adsorption. These will be illustrated by considering the heat of adsorption of hydrogen on tungsten.

In the first place it was found (see fig. 9) that the difference in the heat of adsorption per mol. of hydrogen at $\theta = 0$ and $\theta = 1$ was about 27,000 cal. In these experiments it is probable that the hydrogen is adsorbed with dissociation and, if we assume a fixed interaction energy V_a ergs between atoms adsorbed on neighbouring sites, this gives, for a quadratic lattice,

$$8NV = 27,000J, \tag{3·16}$$

that is, an interaction energy of $2\cdot33 \times 10^{-13}$ erg per pair of adsorbed atoms. The present investigation shows that to account for the total observed change in the heat of adsorption requires an interaction energy V_a ergs between particles adsorbed on neighbouring sites in the final film which is smaller than, and probably considerably smaller than, the above value. It should be remarked here that the value of the interaction energy between neighbouring adsorbed particles in the final film has a marked effect on the sharpness of the change in the gradient in the neighbourhood of $\theta = 0\cdot5$ of the theoretical curve showing the relation between the heat of adsorption and the fraction of sites occupied. The bearing of this on the interpretation of the experimental results will be discussed after we have obtained a lower limit to the value of V_a.

We obtain this lower limit by considering the case in which there are no definite sites for adsorption, but a uniform potential all over the surface.* The behaviour of a one-dimensional film of this type when the particles are always arranged in a configuration of minimum energy† is as follows. Consider r particles arranged in a circle of circumference l. Under the effect of their mutual repulsions the particles will be evenly spaced around the circle and their distance apart may be taken as (Roberts 1938 a)

$$y = l/r.$$

The total interaction energy of the r particles is therefore given by

$$U_r = rY/y^m = Yr^{m+1}/l^m. \qquad (3\cdot17)$$

In the final film, in which n_s sites are occupied, the particles are a distance a apart and the interaction energy between two neighbouring particles is given by

$$V_a = Y/a^m,$$

whence $\qquad\qquad U_r = V_a r^{m+1}/n_s^m. \qquad (3\cdot18)$

It follows that the heat of adsorption is given by

$$q - q_0 = -\frac{dU_r}{dr} = -(m+1)V_a \theta^m. \qquad (3\cdot19)$$

---

* It will be noticed in this connexion that the theory which assumes a fixed interaction energy between particles adsorbed on neighbouring sites gives an upper limit to the value of V_a.

† For a discussion of states of minimum energy see § 3·6.

Thus, for an inverse mth power law of interaction between particles adsorbed on neighbouring sites the total change in the heat of adsorption from $\theta = 0$ to $\theta = 1$ is $(m+1)V_a$.

The behaviour of a two-dimensional film of this type, adsorbed on a quadratic lattice, can be discussed in a similar manner. When r particles are adsorbed on a surface of area $\mathscr{A}$ to form a quadratic lattice, the distance between neighbouring particles is given by

$$y = (\mathscr{A}/r)^{\frac{1}{2}}.$$

The total interaction energy of the film is therefore given by

$$U_r = \frac{2rY}{y^m} = \frac{2Y}{\mathscr{A}^{\frac{1}{2}m}} r^{\frac{1}{2}m+1}, \tag{3.20}$$

and in terms of V_a, the interaction energy between two neighbouring particles in the final film, this can be written as

$$U_r = \frac{2V_a}{n_s^{\frac{1}{2}m}} r^{\frac{1}{2}m+1}, \tag{3.21}$$

whence it follows that the heat of adsorption is given by

$$q - q_0 = -\frac{dU_r}{dr} = -(m+2)V_a \theta^{\frac{1}{2}m}. \tag{3.22}$$

Thus, for an inverse mth power law of interaction between particles adsorbed on a surface to form a quadratic lattice the total change in the heat of adsorption from $\theta = 0$ to $\theta = 1$ is $(m+2)V_a$. For the adsorption of hydrogen on tungsten this leads to a lower limit to the value of V_a given by

$$2(m+2)N(V_a)_{\text{min.}} = 27{,}000J, \tag{3.23}$$

and for an inverse eleventh power law of interaction this gives 7.17×10^{-14} erg per pair of adsorbed atoms for $(V_a)_{\text{min.}}$. This is smaller than the value 2.33×10^{-13} erg per pair of adsorbed atoms given by the fixed interaction theory. Moreover, for a *given* interaction energy between particles adsorbed on neighbouring sites the present theory for an equilibrium film leads to a curve showing the relation between the heat of adsorption and the fraction of sites occupied which is less markedly curved than that obtained from the theory with fixed interaction energy. In both theories the heat curves for an equilibrium film become less markedly curved as the interaction energy between neighbouring adsorbed particles in the complete film is decreased.

It follows that, for an equilibrium film and given initial and final heats of adsorption, the curve showing the variation of the heat of adsorption with the fraction of sites occupied calculated from the present theory is much less markedly curved than that calculated from the theory with fixed interaction energy. On the other hand, the heat curve for an immobile film calculated from the present theory is not very nearly linear as it is in the theory with fixed interaction energy but approaches much more closely the heat curve for a mobile film. Thus, this theory suggests that it is not possible to decide with any certainty, from the approximately linear relation determined experimentally between the heat of adsorption and the fraction of sites occupied, that the hydrogen film adsorbed on tungsten is an immobile film. At the same time, allowing for the variation of the potential energy as in the present treatment, the curves calculated for an immobile film still give a closer representation of the experimental results than do the curves calculated for mobile films. Furthermore, it should be borne in mind that the full curves in figs. 16 and 17 have been calculated for a linear chain, and a more precise conclusion necessarily waits upon similar calculations for a two-dimensional array.*

3·6. States of Minimum Energy

The mutual repulsions between the adsorbed particles tend to make them arrange themselves so that they occupy a state of minimum energy, while the heat motions of the particles tend to make them take up an equilibrium Boltzmann distribution. We can determine the states of minimum energy, and from these we can calculate how the heat of adsorption would vary with the fraction of the surface covered if the adsorbed particles were always arranged in them. The calculations have been carried out for an inverse seventh power law of interaction between neighbouring adsorbed particles.

From $\theta = 0$ to $\theta = \frac{1}{2}$ the particles can be arranged so that no two adjacent sites are occupied. This is shown in fig. 18(a), in which the black dots represent adsorbed particles and the open circles

* It should also be noted that the effect of the assumption that the vibrational partition function is independent of the state of occupation of neighbouring sites has not been determined, and may have a bearing on this question. Cf. § 2·3, footnote.

represent vacant sites for adsorption. From $\theta = 0$ to $\theta = \frac{1}{2}$ the inter-action energy is constant (and zero) and the heat of adsorption is q_0. The adsorption of additional particles at $\theta = \frac{1}{2}$ involves the introduction of pairs of adjacent occupied sites. If a particle is adsorbed on site 3, the particle already on site 4 will migrate to site 5. This is represented schematically in fig. 18(a), in which the vertical arrow represents the adsorption of a particle and the hori-zontal arrow represents the migration of a particle. The result is

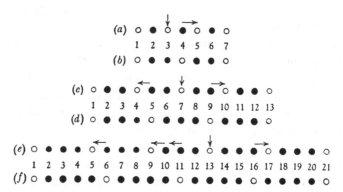

Fig. 18. Arrangement of particles in successive states of minimum energy, showing the way in which the particles rearrange themselves to form the next minimum energy configuration following the adsorption of an addi-tional particle. Black dots represent adsorbed particles, open circles repre-sent vacant sites for adsorption; vertical arrows represent the adsorption of a particle on a vacant site, and horizontal arrows represent the migration of an adsorbed particle from one site to another.

shown in fig. 18(b). Thus between $\theta = \frac{1}{2}$ and $\theta = \frac{2}{3}$ the adsorption of one particle involves the introduction of the interaction energy due to particles adsorbed on two pairs of adjacent sites. From table 2, this interaction energy is $1 \cdot 18 V_a$, and, if q is the heat of adsorption at this stage, then

$$q - q_0 = -1 \cdot 18 V_a.$$

This continues until $\theta = \frac{2}{3}$, at which stage the particles are arranged along the whole row of sites so that adjacent pairs of sites are occupied with vacant sites between these pairs as in fig. 18(c).

If a particle is now adsorbed on site 7, the particle already adsorbed on site 5 migrates to site 4 and that adsorbed on site 9 migrates to site 10, as is shown by the arrows in fig. 18(c) with the result shown in fig. 18(d). Thus the adsorption of one additional particle at $\theta = \frac{2}{3}$ has introduced three groups of particles each consisting of three particles on adjacent sites, in place of four groups of particles each consisting of two particles adsorbed on adjacent sites. Thus, from table 2, at this stage the adsorption of one particle increases the interaction energy by

$$\{3(1\cdot405) - 4(0\cdot589)\}V_a = 1\cdot86\ V_a,$$

which is the value of $q_0 - q$ at $\theta = \frac{2}{3}$. This continues until $\theta = \frac{3}{4}$, at which stage particles are arranged along the whole row of sites so that they are occupied in groups of three adjacent sites with the fourth site vacant as in fig. 18 (e). Thus between $\theta = \frac{2}{3}$ and $\theta = \frac{3}{4}$ the value of $q - q_0$ is $-1\cdot86\ V_a$. This process continues, and the next stage is shown in fig. 18 (e, f), where the horizontal arrows show the migration of particles following the adsorption of a particle on site 13.

At $\theta = (r-1)/r$ the state of minimum energy occurs when groups of $r-1$ adjacent sites are occupied and the rth site is vacant. At $\theta = r/(r+1)$ the state of minimum energy occurs when groups of r adjacent sites are occupied and the $(r+1)$th site is vacant. If at $\theta = (r-1)/r$ an additional particle is adsorbed on one of the vacant sites, a rearrangement of particles involving in all $r^2 - 1$ particles takes place. Following the adsorption of this additional particle, $r+1$ groups, each of which contain $r-1$ particles adsorbed on adjacent sites with one vacant site between each group of occupied sites, rearrange themselves to form r groups, each of which contain r particles adsorbed on adjacent sites with one vacant site separating each group of occupied sites. This rearrangement takes place following the adsorption of one additional particle at all values of θ between $(r-1)/r$ and $r/(r+1)$. If U_r, U_{r-1} respectively are the energies of groups of r, $r-1$ particles adsorbed on adjacent sites, then the change in energy following the adsorption of one additional particle in this range of values of θ is

$$rU_r - (r+1)\ U_{r-1},$$

and the heat of adsorption from $\theta = (r-1)/r$ to $\theta = r/(r+1)$ is therefore given by

$$q - q_0 = (r+1)U_{r-1} - rU_r. \tag{3.24}$$

In this way we obtain the curve consisting of the series of steps

shown in fig. 19. This curve would give the variation of the heat
of adsorption with the fraction of sites occupied if the particles
were always arranged in a minimum energy configuration. The
effect of the heat motions in producing a Boltzmann distribution

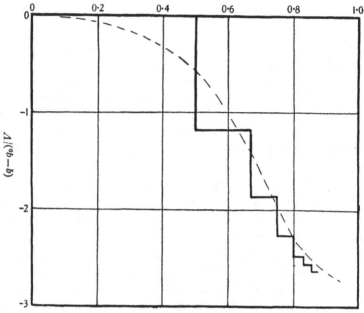

Fig. 19. The variation of the heat of adsorption with the fraction of the surface
covered, when the adsorbed particles are always arranged in a state of
minimum energy, is shown by the full curve for an inverse seventh power
law of interaction between neighbouring adsorbed particles. The broken
curve is reproduced from fig. 16, and shows the variation of the heat of
adsorption when the effect of the thermal motions in producing a Boltzmann distribution of particles over the surface is taken into account. It is
calculated for $\eta = 0.1$.

on the surface is to round off the corners of the steps to give
curve (i) of fig. 16 which is reproduced as the broken curve in
fig. 19. The slight divergence between the step curve and curve (i)
of fig. 16 for high values of θ has already been discussed at the
end of § 3·3.

CHAPTER 4

THE PROCESS OF THE FORMATION OF ADSORBED FILMS

4·1. The Kinetics of the Formation of Immobile Adsorbed Films with Dissociation

It has already been seen that at room temperature the hydrogen film adsorbed on tungsten is probably immobile. We shall begin by considering the formation of films of this type. As an immobile film formed by the dissociation of diatomic gas molecules and the adsorption of the two atoms on neighbouring sites is gradually built up, certain individual surface sites will find themselves surrounded by four filled places. Such sites will not be able to take any part in the adsorption process and will remain bare. The complete film therefore of necessity has gaps or holes in it.* The number of these holes can be determined either empirically, using a model (Roberts 1938 a), or by a statistical method (Roberts and Miller 1939), and is about 8 % of the number of sites on the surface. The presence of the holes must be taken into account in considering the kinetics of the formation of the film.

A lower limit to the number of vacant sites surrounded by occupied sites can be obtained (Roberts and Miller 1939) in the following way. The probability that a given site is vacant is $1 - \theta$, and the probability that in this case each of its z closest neighbours is occupied is from equation (2·19) and table 1 given by

$$\left(\frac{\epsilon_1}{1 + \epsilon_1}\right)^z.$$

Making use of equation (2·24 b), the fraction of the total number of sites which are vacant and are surrounded by four occupied sites is

$$(1 - \theta)\left(\frac{3\theta}{4 - \theta}\right)^4.$$

---

* It may be mentioned here that any immobile film, in which each particle in the gas phase does not on adsorption occupy one site only, will have similar gaps (see Roberts (1938a), Roberts and Miller (1939) and § 6·3), and that such gaps will probably be important as centres of catalytic activity (see Roberts (1935c)).

This function has a maximum value of 0.0646 at $\theta = 0.83$. Thus if an examination is made of a large number of surfaces each of which is 83% filled and the number of vacant sites surrounded by four occupied sites is counted on each, it will be found that the average number of such sites amounts to 6.46% of the total. For immobile films these sites never become available for adsorption. Thus, on the average, no further adsorption can take place when less than 93.5% of the sites are occupied. The final number of occupied sites is *less* than 93.5% because, as the amount adsorbed increases from 83% at which the 6.46% maximum is obtained, some additional vacant sites surrounded by four occupied sites will in general occur. Thus in the final film it may be expected that there will be between 7 and 8% of the sites vacant and not available for adsorption into the first layer. This is in agreement with the value of 8% found by Roberts (1938 a), using an empirical method in which a model of the surface containing 100 numbered sites was used. To eliminate edge effects, the quadratic lattice was pictured as formed of sites arranged on circles wound round a cylinder with the two ends of the cylinder joined together to form a ring. Pairs of sites were filled at random by drawing a numbered card to represent one site and a number 1, 2, 3 or 4 to represent the orientation of the second site with respect to the first site chosen. At each stage, a record was made of the total number of available pairs of vacant sites and of the fractions of these which were surrounded by 6, 5, 4, ..., 0 vacant sites respectively. In this way the former of these quantities can be obtained as a function of the latter. It was found that by the time 92% of the sites were filled there were no longer any pairs of vacant sites available. Thus, from the analysis of this model it is concluded that in an immobile film in which each adsorbed molecule occupies two neighbouring sites there are 8% of vacant sites in the complete film.

For an immobile film it has been seen (§ 2·6) that the statistical parameters are given by

$$\epsilon_0 = \frac{z\theta}{z-\theta}, \quad \epsilon_1 = \frac{z-1}{z} \frac{\theta}{1-\theta}. \tag{4·1}$$

To determine the rate of formation of a film it is necessary first to determine how the rate of condensation varies with θ. For the condensation of particles, each of which occupies a pair of neighbouring sites, this depends on the probability that if a selected site is vacant then a given neighbouring site is also vacant. From table 1 this is simply $1/(1+\epsilon_1)$, so that the rate of condensation per unit area is

$$\frac{a_2 p_2}{(2\pi m_2 kT)^{\frac{1}{2}}} \frac{1-\theta}{1+\epsilon_1}. \tag{4.2}$$

In this equation a_2 is the probability that a molecule condenses when it strikes a place on the surface where two neighbouring sites are vacant, and $p_2/(2\pi m_2 kT)^{\frac{1}{2}}$ is the number of molecules striking unit area per second. Using equation (4.1) we get that the rate of condensation per unit area is

$$\frac{a_2 p_2}{(2\pi m_2 kT)^{\frac{1}{2}}} \frac{z(1-\theta)^2}{z-\theta}. \tag{4.3}$$

It follows that if n_s is the number of sites per unit area, then

$$\frac{d\theta}{dt} = \frac{2z}{n_s} \frac{a_2 p_2}{(2\pi m_2 kT)^{\frac{1}{2}}} \frac{(1-\theta)^2}{z-\theta}, \tag{4.4}$$

and that

$$\frac{2z}{n_s} \frac{a_2 p_2}{(2\pi m_2 kT)^{\frac{1}{2}}} t = \int_0^\theta \frac{z-\theta}{(1-\theta)^2} d\theta,$$

if $\theta = 0$ at $t = 0$. Integrating this we obtain

$$\frac{2z}{n_s} \frac{a_2 p_2}{(2\pi m_2 kT)^{\frac{1}{2}}} t = \frac{z-1}{1-\theta} - (z-1) - \log(1-\theta) = f(\theta), \quad \text{say.} \tag{4.5}$$

To study the formation of hydrogen films on tungsten, Bosworth (1937) used the contact potential method of Langmuir and Kingdon (1925) as developed by Reimann (1935) and applied by Bosworth and Rideal (1937) to the study of sodium films on tungsten. In this method the electron current from an incandescent filament maintained at a fixed temperature to a nearby cold filament is plotted as a function of the voltage between the two filaments, a curve like curve I in fig. 20 being obtained. We shall suppose that curve I refers to a cold tungsten filament with a bare surface. If now an adsorbed film is deposited on the cold surface, the incandescent filament being at such a high temperature that it remains bare, and the experiment is repeated, the whole curve is bodily displaced to II through a distance V equal to the contact potential between the cold bare surface and the cold surface with the adsorbed film on it. The procedure is to determine the calibration curve. Then with the cold surface bare we start with a current, say i_0, corresponding to a point P on the calibration curve. If now adsorption occurs and the current changes by Δi, the contact potential between the bare cold surface and the cold surface with the adsorbed film on it is ΔV. The advantage of the method is that with a short-period recording galvanometer comparatively rapid adsorption processes can be followed.

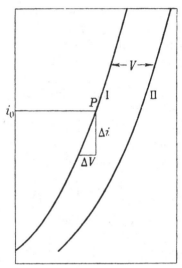

Fig. 20. Schematic diagram showing relation between electron current and potential difference between filaments.

Bosworth found that the contact potential of the tungsten with a hydrogen film on it in its final steady state was 1·04 V. against bare tungsten, the sign being the same as for tungsten with an oxygen film and opposite to that with an alkali metal film. If this corresponds to a film with $\theta = 0 \cdot 92$, we assume that for $\theta = 1$ the contact potential would be 1·12 V. and thus, if V_θ is the contact potential at any stage in building up the film, we assume that $\theta = V_\theta/1 \cdot 12$. Some recent experimental work by Bosworth (1945 a) in which he measured the contact potential difference between a clean tungsten surface and a tungsten wire with an adsorbed layer of nitrogen provides justification for this assumption. Using the $\theta - t$ relation calculated by Roberts (1938 a) empirically, on the basis of a model of the surface, the relation between the contact potential and the fraction of the surface covered shows a maximum deviation of about 3 % from a linear relationship. Bosworth (1945 b) has found similar results for oxygen films adsorbed on tungsten wires, but it should be noted that he quotes a formula (his equation (1)) which is not quite correct (Roberts 1938a; Miller 1947), although it does not affect the numerical results very much. Bosworth (1937) followed the relation between V_θ and the time t as the film built up and thus obtained the relation between θ and t. This can be compared with the relation given by equation (4·5). In fig. 21 the value of the right-hand side of this equation is plotted against the value of t given

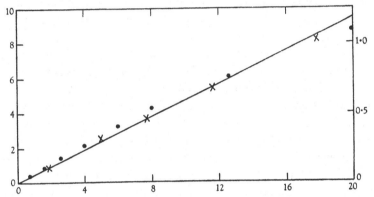

Fig. 21. Relation between $f(\theta)$, from equation (4·5), and t (crosses); and between $\int \dfrac{d\theta}{\phi(\theta)}$, from equation (6·2), and t (dots).

by Bosworth's experiments. The relation is almost exactly linear, and this is what would be expected for an immobile film, while it is far different from that of a mobile film (see § 4·3 below).

4·2. Some Properties of Oxygen Films

We shall now consider some experimental results with oxygen films on tungsten which are consistent with what would be expected if the well-known stable film of oxygen had in it holes of the type discussed in § 4·1. By measuring the heat of adsorption, Roberts (1935b) showed that in addition to the well-known stable film, the heat of adsorption of which is greater than 100 kilocalories per mol., and the amount in which is approximately one atom of oxygen per atom of tungsten, some oxygen is adsorbed with a heat of 40 or 50 kilocalories per mol. This additional oxygen can be removed by heating to a comparatively low temperature (about 1000° K.). Owing to certain complications and difficulties in working with oxygen, which need not be discussed here, it is extremely difficult to obtain an accurate measure of the amount so adsorbed, but it appears to be less than a complete film, and the most likely explanation of its presence is that it consists of molecules held by the bare tungsten atoms in the gaps in the atomic film. The estimated amount so adsorbed agreed with what would be expected on this view, but, although this result is suggestive, it would be unwise, owing to the difficulties already mentioned, to regard it as definite experimental proof of the existence of the holes (see § 6·6).

Van Cleave (1938) has carried out some experiments using the accommodation-coefficient method. The technique was the same as that described in § 1·2 for hydrogen, except that the oxygen was admitted so that it could reach the wire without having to pass through a charcoal tube. Admission of oxygen causes the accommodation coefficient of neon to rise from 0·06, the value for a bare surface, to above 0·3. As the earlier experiments by Roberts had shown, moderate heating of the wire after the deposition of oxygen reduces the accommodation coefficient to a value in the neighbourhood of 0·2, and a low value corresponding to bare tungsten is obtained only after heating to about 2000° K. The actual relation between the accommodation coefficient a measured at *room temperature* and the mean temperature to which the wire with the oxygen film on it had been heated for 1 min. in neon free from oxygen* is shown in fig. 22. The value 0·2 obtained after heating over a considerable range of temperature presumably corresponds to the presence of the well-known stable film. It will be seen that removal of this film begins at temperatures of about 1750° K. Heating the wire to temperatures between about 1500 and 1700° K. caused an *increase* in the value of the accommodation coefficient. This effect is presumably due to some rearrangement in the film, a possible interpretation being that the

---

* There were residual traces of oxygen in the neon and the values plotted were obtained by extrapolation to the time at which the heating current was switched off as described in § 1·2.

film becomes mobile, and that at these temperatures there is an appreciable probability that two gaps should come together and be filled up from residual traces of oxygen in the gas phase. It is evident that this cannot be regarded as direct experimental proof for the existence of holes in the stable film, but it is consistent with the behaviour that would be expected from a film containing such holes.*

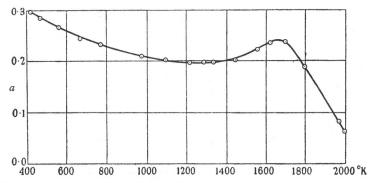

Fig. 22. Accommodation coefficient of a tungsten wire at room temperature as a function of the temperature to which it had been heated for 1 min. in neon containing a trace of oxygen.

4·3. The Kinetics of the Formation of Mobile Films with Dissociation

We shall commence the study of the kinetics of mobile films, in which there is repulsive interaction between adsorbed particles and no two neighbours of a given site are neighbours of each other, by considering the states of lowest energy as we did in § 2·2 in connexion with heats of adsorption. When $\theta = 0·5$ the particles will be arranged as shown in fig. 10, that is, no two neighbouring sites will be vacant and no further adsorption of diatomic molecules with dissociation will be possible. The curve showing the relation between θ and the time will obviously be of the type marked $\eta = 0$ in fig. 23, the shape of the earlier part being calculated as shown below.

When we take into account the effect of the thermal motion of

* It may be mentioned here that Van Cleave found evidence that oxygen can be adsorbed on the top of the first film and not in the gaps only. This is a further indication of the complicated nature of the processes involved in connexion with the adsorption of oxygen (see Chapter 6).

the particles in making configurations other than that of lowest energy occur, it is evident from the above considerations that there will be a rapid change in the rate of formation of the film in the neighbourhood of $\theta = 0.5$. The reason for this is that, for $\theta \geqslant 0.5$, configurations with two neighbouring vacant sites cannot occur without causing an increase in the number of neighbouring sites occupied, i.e. an increase in the interaction energy of the adsorbed

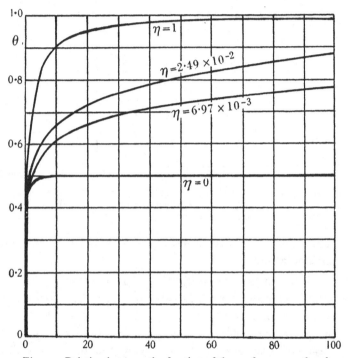

Fig. 23. Relation between the fraction of the surface covered and the time for the formation of a mobile film.

particles. For example, in fig. 10, if the particle on a site which we shall call site 1 moves to a neighbouring site, adsorption can occur on site 1. Since, by Boltzmann's law, the occurrence of such configurations is relatively improbable, the rate of condensation becomes slow.

We obtain the formula for the rate of condensation as follows (Roberts 1937). If the central site is vacant, the chance that site 1 is vacant is, by expression (2·9), given by $1/(1 + \epsilon)$. The probability that the central site is vacant is $1 - \theta$. Thus the probability that a molecule striking the surface at a given place finds two neighbouring sites vacant is $(1 - \theta)/(1 + \epsilon)$. The rate of condensation of molecules per unit area is

$$\frac{a_2 p_2}{(2\pi m_2 kT)^{\frac{1}{2}}} \frac{1 - \theta}{1 + \epsilon}, \qquad (4\cdot6)$$

where, as before, a_2 is the probability that a molecule condenses when it strikes the surface at a place where two neighbouring sites are vacant. Thus, if there is no re-evaporation, we have, as in equation $(4\cdot4)$,

$$\frac{d\theta}{dt} = \frac{2}{n_s} \frac{a_2 p_2}{(2\pi m_2 kT)^{\frac{1}{2}}} \frac{1 - \theta}{1 + \epsilon}, \qquad (4\cdot7)$$

where n_s is the number of sites per unit area. Integrating this equation we obtain

$$\frac{2}{n_s} \frac{a_2 p_2}{(2\pi m_2 kT)^{\frac{1}{2}}} t = \int_0^\theta \frac{1 + \epsilon}{1 - \theta} d\theta, \qquad (4\cdot8)$$

if $\theta = 0$ at $t = 0$. By using the value of ϵ given by equation $(2\cdot10)$ this equation can be integrated numerically. At constant pressure and temperature it shows that t is proportional to $\int_0^\theta \frac{1 + \epsilon}{1 - \theta} d\theta$, so that the variation of θ with t is given by plotting θ as a function of $\int_0^\theta \frac{1 + \epsilon}{1 - \theta} d\theta$. This is done in fig. 23 for various values of $\eta = e^{-V/kT}$.

If $\eta = 1$, i.e. the interaction energy between neighbouring adsorbed atoms is negligibly small, we have a random distribution, and the probability of finding two neighbouring sites vacant becomes $(1 - \theta)^2$. If we write $(1 - \theta)^2$ instead of $(1 - \theta)/(1 + \epsilon)$, equation $(4\cdot8)$ becomes

$$\frac{2}{n_s} \frac{a_2 p_2}{(2\pi m_2 kT)^{\frac{1}{2}}} t = \int_0^\theta \frac{d\theta}{(1 - \theta)^2}, \qquad (4\cdot9)$$

and this yields

$$\frac{2}{n_s} \frac{a_2 p_2}{(2\pi m_2 kT)^{\frac{1}{2}}} t = \frac{\theta}{1 - \theta}. \qquad (4\cdot10)$$

It is interesting to compare this equation for condensation into a film in which there is a random distribution of single occupied sites with equation $(4\cdot5)$,

$$\frac{2z}{n_s} \frac{a_2 p_2}{(2\pi m_2 kT)^{\frac{1}{2}}} t = \frac{z - 1}{1 - \theta} - (z - 1) - \log (1 - \theta),$$

which refers to a film in which there is a random distribution of pairs of occupied sites. We suppose that a_2, p_2, n_s, m_2 and T are the same in the two cases. The relative times taken for a fraction

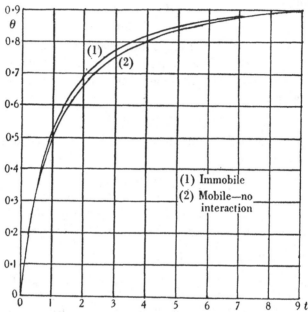

Fig. 24. Rate of formation of (1) immobile film and (2) mobile film with no interaction.

θ of the sites to be occupied are then proportional to the right-hand members of these two equations; θ is plotted as a function of these two expressions in fig. 24. It will be seen that the two curves are very similar and that they cross just below $\theta = 0.9$. The reason for this is that the curve for the immobile film approaches a limiting value of $\theta = 0.92$, while the other approaches a limiting value of $\theta = 1$.

CHAPTER 5

EVAPORATION PROCESSES AND THE PRODUCTION OF ATOMIC HYDROGEN

5·1. The Production of Atomic Hydrogen by Hot Tungsten

When a tungsten filament is heated to a sufficiently high temperature in hydrogen contained in a glass vessel the walls of which are cooled in liquid air, Langmuir (1912, 1915, 1926) showed that there is a continuous diminution of pressure and interpreted this as being due to the production by the hot tungsten of atomic hydrogen which is adsorbed on the surface of the glass. In order to investigate the rate of production quantitatively it is essential to use an efficient method of trapping the atoms, since, if it can be assumed that any atom produced is trapped before it combines with another atom, the rate of disappearance of gas, obtained from the volume of the vessel and the rate of diminution of pressure, gives a measure of the rate of production of atoms. Langmuir's measurements themselves showed that glass cooled in liquid air was not consistently efficient in this way, since the rate of production of atoms under definite conditions was not reproducible.

Atomic hydrogen is known to react with molybdenum oxide. In order to trap the atoms efficiently, and obtain accurate measurements of their rate of production, Bryce (1936) deposited this oxide to a depth of more than fifty layers on the walls of the vessel containing the tungsten by heating a subsidiary molybdenum filament in oxygen. This oxygen was pumped off and the experiment on the production of atomic hydrogen was begun, the vessel containing the filament being immersed in melting ice to keep the temperature constant. The results obtained showed that the earlier estimates of the rate of production under given conditions were too low by a factor at least as great as two hundred. From this, and the fact that subsidiary experiments showed that there was no fatigue effect in the molybdenum oxide, it can be concluded that the oxide is a much more reliable trap than the glass surface cooled in liquid air; the considerations given below indicate that it must provide almost a completely efficient trap.

With the filament at a fixed temperature the effect of varying the pressure on the rate of production of atoms was determined. The results showed that at constant filament temperature the rate of production is inversely proportional to the square root of the hydrogen pressure.

The relation between the rate of production and the temperature of the filament was determined accurately. After correcting for the distribution of temperature along the filament, the results showed that n, the number of hydrogen atoms produced per cm.^{2} per sec., is given by

$$n = 2 \cdot 5 \times 10^{24}\, p_{mm.}\, e^{-45,000/RT}, \qquad (5 \cdot 1)$$

where $p_{mm.}$ is the pressure in mm. of mercury, R is the gas constant, T° K. is the temperature of the tungsten and the gas is at 0° C. The experimental results on which this equation is based cover a temperature range from 1148 to 1420° K., and a pressure range from 3×10^{-3} to $3 \cdot 7 \times 10^{-2}$ mm. of mercury.

It is evident that to explain these results we must consider in some detail the various processes occurring at the surface, including the evaporation of adsorbed particles from it.

5·2. Groups of Processes occurring at the Surface

In the case of adsorption without dissociation there are only two processes to be taken into account: (i) condensation of molecules, and (ii) evaporation of molecules. When each molecule occupies one site on the surface and when there is no interaction, we have seen in § 2·1 that the condition of balance between these two processes leads to the Langmuir adsorption isotherm.

When dissociation occurs on adsorption there are more possible processes at the surface to be considered, and in discussing the equilibrium problem we must also take into account the fact that at a given temperature and pressure the gas in its equilibrium state contains both undissociated and dissociated molecules. The equilibrium condition for the reaction

$$H_2 = 2H$$

in the gas phase is $$p_1 = 10^3 K^{\frac{1}{2}} p_2^{\frac{1}{2}}, \qquad (5 \cdot 2)$$

where p_1 and p_2 are the partial pressures in dyne cm.$^{-2}$ of atomic and molecular hydrogen respectively and K is the equilibrium constant with pressures measured in atmospheres. Although for preciseness the case of hydrogen is considered, the treatment applies equally well to any diatomic molecule.

Suppose the gas is contained in a tungsten box maintained at the required temperature T, the surface of the metal being free from oxygen and other adsorbed impurities. Quite apart from any reactions taking place in the gas phase, the processes occurring at the surface must themselves be able to set up and maintain the degree of dissociation of the hydrogen corresponding to the temperature of the walls and the pressure in the box. There are three pairs of surface processes to be considered and, according to the principle of detailed balancing, each pair of these must balance individually. These processes are:

(i) (a) Evaporation of atoms from the adsorbed film. (b) An atom from the gas strikes the surface where there is a vacant site and condenses.

(ii) (a) Two neighbouring adsorbed atoms combine and evaporate as a molecule. (b) A molecule strikes the surface where two neighbouring sites are vacant. dissociates and the atoms are adsorbed.

(iii) (a) A gas atom strikes an adsorbed atom, combines with it and the two evaporate as a molecule. (b) A gas molecule strikes the surface where there is one vacant site or more, one atom is adsorbed, the other goes into the gas phase.

Neglecting the interactions between adsorbed atoms, Roberts (1936) investigated the balance between these processes, and Roberts and Bryce (1936) gave the theory of the production of atomic hydrogen. Subsequently, Roberts (1937) showed how to take account of the interactions between the adsorbed atoms and showed that their inclusion produced no important difference. This theory we now consider, but first it is desirable to discuss some of the fundamental assumptions on which any such theory must be based.

5·3. Fundamental Assumptions

The adsorbed particles can be regarded as oscillators moving in various quantum states in the field at the surface of the solid. Lennard-Jones and Devonshire (1936) have discussed the excitation of such oscillators and the evaporation of the particles by inter-

action with the elastic waves of the solid. This treatment super-
sedes completely all earlier attempts to obtain formulae for
evaporation based on a classical picture. They have shown how
calculations can be carried out for the case in which the oscillations are normal to the surface, so that the state can be specified
by one quantum number, and when only one quantum of
thermal vibration of the solid is transferred at a time. This
practically restricts the treatment to atoms or molecules held by
van der Waals forces, since only these would be likely to be
evaporated by one such quantum.* They assume that the potential
energy between the adsorbed particles and the surface can be
represented by a Morse function (see § 1·1). The interesting result
has been obtained that even at very low temperatures evaporation
in two stages is relatively very frequent. The two stages involved
are, first, excitation to a higher vibrational level and then com-
munication from the solid of a further quantum sufficient to cause
evaporation. This mechanism is illustrated by fig. 25, in which
the dots give the relative populations of the various states, the
crosses give the relative probability of evaporation from each state
and the broken crosses, which are drawn to a different scale, give
the product of the relative populations and the probability of
evaporation, that is, the relative numbers of evaporating particles
coming from the different states. The calculations were made for
$T = 300°$ K. for a solid with a characteristic temperature $\Theta =
510°$ K., and for a heat of adsorption from the lowest state of
1000 calories per mol. It will be seen that evaporating particles
are most likely to come from the third state, and that more come
from the fifth state than from the ground state, in spite of the
relatively low populations of the third and fifth states.

Lennard-Jones and Devonshire analyse the assumption that for
a given desorption process the rate of evaporation of particles per
unit area is proportional to

$$A\theta e^{-\chi/kT}, \tag{5·3}$$

---

* Strachan (1937) has later shown that the probability of the simultaneous
transfer of several quanta of energy by the solid is small. Lennard-Jones and
Goodwin (1937) have therefore examined the possibility that the energy necessary
to excite and evaporate adsorbed atoms held by chemical forces may come from
the metallic electrons. It has not been possible to carry the calculations in this
case quite as far as in that considered by Lennard-Jones and Devonshire.

where χ is the energy of desorption per molecule from the lowest state and A is a constant. In this formula, as in their theory, it is tacitly assumed that there is no interaction between adsorbed particles. They carry out calculations for particles of molecular weight 2 and for a heat of desorption of 1000 calories per mol. and

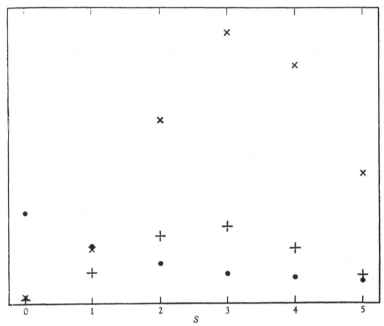

Fig. 25.   The probability of evaporation of an adsorbed molecule from various quantum states of vibration perpendicular to the surface of a solid. The dots (●) give the relative population of the various states; the crosses (×) give the relative probability of evaporation from each state; and the broken crosses (⊹) (which are on a different scale) give the product of the relative populations and the probability of evaporation. ($T = 300°$ K., $\Theta = 510°$ K.)

$\Theta = 510°$ K. at $T = 30$ and $T = 300°$ K., and show that A is not constant but that $$A_{300}/A_{30} \approx 6.$$

The essential thing from our point of view is that their calculation justifies the use of formula (5·3) because over this range of temperature and for the given value of χ the term $e^{-\chi/kT}$ varies by a factor of about 10^7, so that the correct formula with A a function of T and formula (5·3) with A constant and a slight adjustment in the value of χ would be completely indistinguishable in any experimental investigation.

We shall make a similar assumption throughout the subsequent treatment, and shall suppose that, when there is an interaction energy V (positive if the force is repulsive) between particles adsorbed on neighbouring sites, the relative probability of evaporation of a particle on a site with $\theta_1 + \ldots + \theta_z$ neighbouring sites occupied contains a factor

$$\exp\left[-\{\chi - (\theta_1 + \ldots + \theta_z)V\}/kT\right],$$

and that the factor which corresponds to A in equation (5·3) does not vary appreciably with T or with $\theta_1 + \ldots + \theta_z$. Although A is undoubtedly in fact a function of T and of $\theta_1 + \ldots + \theta_z$, the essential point is that in the important temperature range

$$\{\chi - (\theta_1 + \ldots + \theta_z)V\}/kT$$

is assumed to be sufficiently large for variations in the exponential term occasioned by changes in $\theta_1 + \ldots + \theta_z$ or in T to swamp completely the accompanying changes in A. Lennard-Jones and Devonshire have, as we have seen, justified the assumption of an effectively constant A at different temperatures for van der Waals adsorption and for no interaction between adsorbed particles. We extend it to cover the effect of interaction between adsorbed particles and processes in which the heat of adsorption indicates that chemisorption is involved.

Lennard-Jones and Devonshire have also examined the theory of condensation for a similar model, and have shown that for a molecule striking a vacant site the condensation coefficient varies little with temperature, being 0·081 at 30° K. and 0·103 at 300° K.

5·4. Theory of Equilibrium and of Production of Atomic Hydrogen including Effect of Interactions

We now consider the balance between the three pairs of processes given in § 5·2, taking into account interactions between the adsorbed atoms (Roberts 1937). We assume that each site has z closest neighbours, that no two neighbours of a given site are neighbours of each other, and that there is an interaction energy V between atoms adsorbed on neighbouring sites and no appreciable interaction energy if they are at a greater distance. The relative probability of a given arrangement of atoms on the central site o and its neighbours $1, \ldots, z$ specified as before by the numbers $\theta_0, \theta_1, \ldots, \theta_z$ is given by equation $(2 \cdot 9)$.

Let us first consider the pair of processes (i). If we write $\eta = e^{-V/kT}$ the expression for the rate of evaporation of atoms from sites surrounded by $(\theta_1 + \ldots + \theta_z)$ occupied sites will, as discussed in the preceding section, contain a factor

$$\eta^{-(\theta_1 + \ldots + \theta_z)}.$$

Since we are discussing the evaporation of an atom from the central site, we consider only arrangements in which the central site is occupied, i.e. $\theta_0 = 1$. The relative probability of an assigned set of values of $\theta_1, \ldots, \theta_z$ is, by equation $(2 \cdot 9)$, given by

$$(\eta \epsilon)^{\theta_1 + \ldots + \theta_z},$$

so that the rate of evaporation from unit area of the surface is given by

$$B_1 \theta \frac{\Sigma (\eta \epsilon)^{\theta_1 + \ldots + \theta_z} \eta^{-(\theta_1 + \ldots + \theta_z)}}{\Sigma (\eta \epsilon)^{\theta_1 + \ldots + \theta_z}}$$

or by

$$B_1 \theta \frac{\Sigma \epsilon^{\theta_1 + \ldots + \theta_z}}{\Sigma (\eta \epsilon)^{\theta_1 + \ldots + \theta_z}},$$

where $B_1 \theta$ would be the rate of evaporation of atoms per unit area for a given θ if the atoms were arranged in such a way that there were no interactions. Since the sums are to be taken for all possible combinations of the various values of $\theta_1, \ldots, \theta_z$, this becomes

$$B_1 \theta \left(\frac{1 + \epsilon}{1 + \eta \epsilon} \right)^z. \tag{5.4}$$

The rate of the inverse condensation process per unit area is

$$a_1 \frac{p_1}{\mu_1} (1 - \theta), \tag{5.5}$$

where it is assumed that a_1, the probability that an atom condenses when it strikes a vacant site, is independent of the state of occupation of neighbouring sites. For the balance of processes (i), we have, then,

$$\frac{\theta}{1 - \theta} = \frac{a_1 p_1}{B_1 \mu_1} \left(\frac{1 + \eta \epsilon}{1 + \epsilon} \right)^z. \tag{5.6}$$

Now consider (ii), the evaporation and condensation of molecules.

We have already seen in equation (4·6) that the rate of condensation of molecules on unit area per second is

$$\frac{a_2 p_2}{\mu_2} \frac{1-\theta}{1+\epsilon}. \tag{5·7}$$

In the inverse evaporation process atoms on sites o and 1 combine and evaporate. The probability that site o is occupied is θ. If site o is occupied, the probability that site 1 is also occupied is, from (2·9),

$$\frac{\Sigma(\eta\epsilon)^{1+\theta_2+\dots+\theta_z}}{\Sigma(\eta\epsilon)^{\theta_1+\theta_2+\dots+\theta_z}} = \frac{\eta\epsilon}{1+\eta\epsilon}.$$

That is, the probability that sites o and 1 are occupied together is

$$\theta \frac{\eta\epsilon}{1+\eta\epsilon}.$$

The rate of evaporation will contain a factor depending on the state of occupation of the sites surrounding o and 1. We continue to specify the occupation of the sites around o by $\theta_1, \dots, \theta_z$ and specify the occupation of those round 1 by $\theta_1', \dots, \theta_z'$ with $\theta_1' = \theta_0$. No other dashed site coincides with an undashed site. From considerations exactly similar to those leading to expression (5·4) the average rate of evaporation of molecules per unit area of surface is

$$B_2 \theta \frac{\eta\epsilon}{1+\eta\epsilon} \frac{\Sigma(\eta\epsilon)^{1+\theta_2+\dots+\theta_z} \eta^{-(\theta_2+\dots+\theta_z)} (\eta\epsilon)^{1+\theta_2'+\dots+\theta_z'} \eta^{-(\theta_2'+\dots+\theta_z')}}{\Sigma(\eta\epsilon)^{1+\theta_2+\dots+\theta_z} (\eta\epsilon)^{1+\theta_2'+\dots+\theta_z'}},$$

where the sums are to be taken for all possible combinations* of values of $\theta_2, \dots, \theta_z, \theta_2', \dots, \theta_z'$ and B_2 is the rate of evaporation of molecules from unit area of a complete film. This expression is equal to

$$B_2 \theta \frac{\eta\epsilon}{1+\eta\epsilon} \left(\frac{1+\epsilon}{1+\eta\epsilon}\right)^{2z-2}. \tag{5·8}$$

For the condition of balance of processes (ii) obtained by equating (5·7) and (5·8) we have, since, by (2·10),

$$\frac{\theta}{1-\theta} = \frac{\epsilon(1+\eta\epsilon)}{1+\epsilon},$$

that

$$\frac{\theta}{1-\theta} = \left(\frac{a_2 p_2}{B_2 \mu_2}\right)^{\frac{1}{2}} \frac{1}{\eta^{\frac{1}{2}}} \left(\frac{1+\eta\epsilon}{1+\epsilon}\right)^z. \tag{5·9}$$

* It may be mentioned that θ_2', etc., are not all necessarily independent of θ_2, etc., but with the present method of averaging this interdependence must be neglected (see § 5·6).

We now consider the third pair of processes. When a hydrogen atom from the gas phase strikes an adsorbed atom and the two combine and evaporate as a molecule, there is a considerable evolution of heat. We assume that the frequency of occurrence of this process per unit area is given by

$$\frac{B_3 p_1 \theta}{\mu_1}, \tag{5·10}$$

where B_3, which is the probability that a gas atom striking an adsorbed atom combines with it, is to a first approximation independent of the temperature and of the state of occupation of surrounding sites. In the inverse process, when a molecule from the gas phase strikes an unoccupied site (the central site) and one atom is adsorbed and the other escapes to the gas phase, the energy *absorbed* is

$$q_d - (q_1)_0 + V(\theta_1 + \dots + \theta_z), \tag{5·11}$$

where q_d is the energy of dissociation per molecule in the gas phase and $(q_1)_0$ the heat of adsorption of an atom on a bare surface. The frequency of occurrence of this process is therefore taken as containing a factor $\eta^{\theta_1 + \dots + \theta_z}$. If the central site is vacant, the relative probability of a configuration of given $\theta_1, \dots, \theta_z$ is by (2·9) given by $\epsilon^{\theta_1 + \dots + \theta_z}$. Thus the frequency of occurrence of this process per unit area per second is

$$\frac{B_4 p_2}{\mu_2} (1 - \theta) \frac{\Sigma(\eta\epsilon)^{\theta_1 + \dots + \theta_z}}{\Sigma \epsilon^{\theta_1 + \dots + \theta_z}},$$

where B_4 is the probability that a molecule, which strikes a place on the surface where there is a vacant site, dissociates so that one atom is adsorbed and the other escapes into the gas phase. This expression is equal to

$$\frac{B_4 p_2}{\mu_2} (1 - \theta) \left(\frac{1 + \eta\epsilon}{1 + \epsilon} \right)^z. \tag{5·12}$$

For the equilibrium of the pair of processes (iii) we obtain from (5·10) and (5·12)

$$\frac{\theta}{1 - \theta} = \frac{B_4}{B_3} \left(\frac{m_1}{m_2} \right)^{\frac{1}{2}} \frac{p_2}{p_1} \left(\frac{1 + \eta\epsilon}{1 + \epsilon} \right)^z. \tag{5·13}$$

Substituting the value of p_1 given by equation (5·2) in (5·6) and (5·13), the three equations (5·6), (5·9) and (5·13) which give $\theta/(1-\theta)$ become respectively

$$\frac{\theta}{1-\theta} = \frac{10^3 K^{\frac{1}{4}}}{\mu_1} \frac{a_1}{B_1} p_2^{\frac{1}{2}} \left(\frac{1+\eta\epsilon}{1+\epsilon}\right)^z, \qquad (5·14)$$

$$\frac{\theta}{1-\theta} = \left(\frac{1}{\mu_2\eta}\right)^{\frac{1}{2}} \left(\frac{a_2}{B_2}\right)^{\frac{1}{2}} p_2^{\frac{1}{2}} \left(\frac{1+\eta\epsilon}{1+\epsilon}\right)^z, \qquad (5·15)$$

$$\frac{\theta}{1-\theta} = \frac{1}{10^3 K^{\frac{1}{4}}} \left(\frac{m_1}{m_2}\right)^{\frac{1}{4}} \frac{B_4}{B_3} p_2^{\frac{1}{2}} \left(\frac{1+\eta\epsilon}{1+\epsilon}\right)^z. \qquad (5·16)$$

These are all of the form

$$\frac{\theta}{1-\theta} = B p_2^{\frac{1}{2}} \left(\frac{1+\eta\epsilon}{1+\epsilon}\right)^z. \qquad (5·17)$$

This isotherm will be discussed in § 5·6. The different expressions for B must all be equal, which gives us relations between a_1, B_1, a_2, B_2, B_3 and B_4.

It should be noted that if there are no interactions then η is unity and these equations reduce to

$$\frac{\theta}{1-\theta} = \frac{10^3 K^{\frac{1}{4}}}{\mu_1} \frac{a_1}{B_1} p_2^{\frac{1}{2}}, \qquad (5·18)$$

$$\frac{\theta}{1-\theta} = \left(\frac{a_2}{\mu_2 B_2}\right)^{\frac{1}{2}} p_2^{\frac{1}{2}}, \qquad (5·19)$$

$$\frac{\theta}{1-\theta} = \frac{1}{10^3 K^{\frac{1}{4}}} \left(\frac{m_1}{m_2}\right)^{\frac{1}{4}} \frac{B_4}{B_3} p_2^{\frac{1}{2}}, \qquad (5·20)$$

each of which is of the form

$$\frac{\theta}{1-\theta} = A p_2^{\frac{1}{2}}, \qquad (5·21)$$

where A depends only on the temperature.

These results will now be applied to the production of atomic hydrogen. Atoms in the gas phase are due to two of the processes which have been discussed above. These are (i) (a), in which atoms evaporate from the adsorbed film, and (iii) (b), in which a gas molecule strikes a bare surface and dissociates, one atom being adsorbed and the other escaping into the gas phase. Since the rate of production of hydrogen atoms at a given temperature is proportional to the square root of the pressure, then either the film is very sparsely occupied and process (i) (a) is the important one, or the film is nearly complete and process (iii) (b) is the important one. The rate of evaporation of atoms from unit area of the surface (process (i) (a)) is

$$B_1\theta\left(\frac{1+\epsilon}{1+\eta\epsilon}\right)^z.\tag{5.4}$$

Substituting the value of ϵ given by (5.14) in this and using the fact that θ is small, it becomes

$$\frac{10^3}{\mu_1}K^{\frac{1}{2}}a_1p_2^{\frac{1}{2}}.\tag{5.22}$$

The variation of μ_1 with temperature, which is as $T^{\frac{1}{2}}$, is negligible compared with that of K which is an exponential function of temperature. The adsorption of a hydrogen atom on tungsten is a highly exothermic process (Roberts 1935 b), and it would not be expected that a_1 would vary much with temperature, a conclusion which is supported by the calculations of Lennard-Jones and Devonshire referred to in § 5.3. Treating a_1 as constant, the temperature variation of the rate of production of atomic hydrogen is due entirely to the temperature variation of $K^{\frac{1}{2}}$.

The rate of production of atomic hydrogen by process (iii) (b) is

$$\frac{B_4p_2}{\mu_2}(1-\theta)\left(\frac{1+\eta\epsilon}{1+\epsilon}\right)^z.\tag{5.12}$$

Substituting the value of ϵ given by (5.16) in this and using the fact that θ is nearly unity in this case, it becomes

$$\frac{10^3B_3}{\mu_1}K^{\frac{1}{2}}p_2^{\frac{1}{2}}.\tag{5.23}$$

As in the other case the variation of μ_1 with temperature can be neglected compared with that of K. B_3 is the probability that when an atom from the gas phase strikes an adsorbed atom, the two combine and evaporate as a molecule. A large amount of heat is evolved when this takes place, and it can be assumed that B_3 is to a first approximation independent of temperature. Thus, the temperature variation of the rate of production of atomic hydrogen is due entirely to the temperature variation of $K^{\frac{1}{2}}$. Thus a determination of the temperature variation does not make it possible to decide between the two processes.*

* The same result follows if the interactions are neglected (Roberts and Bryce 1936) as can be seen by considering equations (5.18)–(5.21).

We shall now show that this theoretical temperature variation agrees with that determined experimentally. Values of $\log_{10} K$ can be calculated from the data given by Giauque (1930), but using the value 101,000 cal. (obtained from spectroscopic data) for the heat of dissociation, instead of the older value 102,800 used by him. If values of $\log_{10} K^{\frac{1}{2}}$ deduced from these data are plotted against $1/T$ a linear relation is obtained over the required temperature range and, using values of $K^{\frac{1}{2}}$ from this plot, the relative rates of production of atomic hydrogen at the tungsten surface at various temperatures can be obtained. The results of this calculation together with the measured rates are given in table 4. The agreement is satisfactory.

TABLE 4

Temperature ° K.	Rate of production of H atoms per cm.² per sec. at $= 1.25 \times 10^{-2}$ mm. mercury	
	Observed	Theoretical
1420	316×10^{14}	(316×10^{14})
1338	115×10^{14}	103×10^{14}
1243	32×10^{14}	22×10^{14}
1148	7×10^{14}	4×10^{14}

In order to decide which of the two processes is the important one, it is necessary to determine experimentally whether the surface is sparsely or nearly fully covered in the above experiments. This has been done by Bosworth (1937), using the method of contact potentials described in § 4·1. He has shown that, under the conditions obtaining in Bryce's experiments, the covering would always be greater than 0·6 or 0·7. We therefore conclude that the second process considered above, in which a molecule strikes a bare place on the surface, one atom being adsorbed and the other passing into the gas phase, is under the conditions of Bryce's experiments the one primarily responsible for the production of atomic hydrogen.

5·5. True and Apparent Heats of Evaporation of Adsorbed Films

In a number of experiments, particularly with oxygen on tungsten, the heat of adsorption has been deduced from measurements of the relative rates of evaporation of adsorbed films at different temperatures. Various electrical methods have been used to measure the amount of oxygen on the surface.

Langmuir and Kingdon (1925) showed that an adsorbed film of caesium on tungsten or on tungsten already covered with oxygen increases the electron emission at a given temperature enormously. Further, when there is an oxygen film on the tungsten, the caesium film is much more stable than on bare tungsten. The effect of this is evident if a filament is heated to a temperature in the neighbourhood of about 800–1000° K. in the presence of caesium vapour at a pressure corresponding to the vapour pressure of caesium at about 20° C. Under these conditions an oxygen-covered filament has much more caesium on it than a bare tungsten filament, and the emission from it is much greater. Langmuir and Villars (1931) have made this fact the basis of a method of measuring the amount of oxygen on a tungsten surface. In applying this method to determine relative rates of evaporation of oxygen they allowed the evaporation to proceed for a measured time at a high temperature (1856–2070° K.). They then cooled the filament down to a temperature in the range 800–1000° K. in the presence of caesium vapour and measured the emission. Assuming that a given emission under definite conditions corresponds to a given fraction of surface covered with oxygen, they were able to deduce relative rates of evaporation at different temperatures.

Johnson and Vick (1935) measured the thermionic emission at a fixed high temperature from a tungsten filament and studied directly the effect of an oxygen film on the thermionic current. As the film evaporated, the emission increased and a cathode-ray oscillograph was used to follow the changes. Their system had the advantage of involving only tungsten and oxygen. Bosworth and Rideal (1937) have given an account of similar experiments in which the contact potential method described in § 4·1 was used.

The heat of adsorption per particle, which we shall for convenience call the apparent heat of evaporation q_a, is in all these experiments deduced from the rate of evaporation r by means of the formula

$$q_a = -k \frac{d \log r}{d(1/T)},$$

(5·24)

where k is Boltzmann's constant. Langmuir and Villars found an apparent heat of 164,000 calories per mol. for small values of θ, and state that the value becomes less as θ increases but give no details. Johnson and Vick found 147,000 for small θ. Bosworth and Rideal found 150,000 for small θ, diminishing to about 70,000 as θ increases.

In this section we will discuss the meaning of the heats deduced in this way when there is interaction between the adsorbed particles. For simple adsorption, in which each particle in the gas phase occupies one site for adsorption, the true heat of adsorption is obtained, but Roberts (1938a) has shown that in other cases this is not generally true, and to interpret the results correctly we must know the nature of the desorption process involved. To illustrate this we shall discuss the relevant physical processes involved in connexion with the adsorption of dimer* molecules as has been considered by Roberts (1938a) and Miller (1947).

First, we shall consider the case in which adsorption takes place with dissociation. In the evaporation from the film the only important process is the recombination of adsorbed atoms and the subsequent evaporation of molecules.

Consider the true heat of adsorption when $\theta = 0$. A gas molecule strikes the surface and the two atoms are adsorbed on two neighbouring sites. The energy of the system is diminished by χ_2, where χ_2 is the heat of adsorption when the two atoms remain on neighbouring sites and all the surrounding sites are vacant. In the normal state of the system with only two atoms on the surface they are on neighbouring sites for only a very small fraction of the total time. In passing to this normal state there is a further diminution of the potential energy of the system by V, where V is the interaction energy between atoms adsorbed on neighbouring sites. The true heat of adsorption $(q_2)_0$ is therefore given by

$$(q_2)_0 = \chi_2 + V.$$

When $\theta = 0$ we may consider in the following way the apparent heat of evaporation deduced from the temperature variation of the rate of evaporation. In the limit there are just two atoms left on

* By a dimer molecule is meant a molecule such as the hydrogen molecule, which if it is adsorbed as a molecule must be regarded as occupying two adjacent sites on the surface.

the surface. Except for a very small fraction of the total time they will not be on neighbouring sites. In order to evaporate as a molecule they must occupy neighbouring sites, and when they do so the potential energy of the system is increased by V. When they evaporate, the potential energy* is further increased by χ_2 so that the heat factor in the exponential term in the expression for the rate of evaporation is $\chi_2 + V$, and this is the apparent heat of evaporation obtained from the variation of the rate with temperature. We have therefore, for $\theta = 0$,

$$(q_a)_0 = (q_2)_0 = \chi_2 + V. \qquad (5 \cdot 25\,a)$$

When $\theta = 1$ the following considerations apply to the true heat of adsorption. In the limit there are two vacant sites left on the surface. Except for a very small fraction of the total time these will not be neighbours. When they are neighbours, the potential energy is greater by V than it is in the normal state. When a molecule from the gas phase comes and occupies these two sites, the energy of the system is diminished by $\chi_2 - (2z - 2)V$. The total change in energy in going from the normal state with two vacant sites to the final state with all sites occupied is therefore $\chi_2 - (2z - 2)V - V$, so that $(q_2)_1$, the true heat of adsorption with $\theta = 1$, is, for $z = 4$, given by

$$(q_2)_1 = \chi_2 - 7V. \qquad (5 \cdot 26\,a)$$

Since the surface is fully covered, no reorganization is necessary before evaporation can occur, and therefore the heat factor in the exponential term in the expression for the rate of evaporation is $\chi_2 - (2z - 2)V$, which is the energy that a molecule must acquire† in order to leave the surface. This is the apparent heat of evaporation deduced from the variation of the rate of evaporation with temperature. Thus, for $z = 4$, we have

$$(q_a)_1 = \chi_2 - 6V. \qquad (5 \cdot 27\,a)$$

When adsorption occurs without dissociation,‡ and each molecule occupies two closest neighbour sites, then, if a molecule con-

---

* We may in connexion with these simple physical considerations neglect the contribution of changes in kinetic energy to heats of adsorption.

† Apart from any heat of activation in the adsorption process, see Roberts (1937).

‡ See Miller (1947).

denses on a bare surface, the energy of the system is decreased by χ, where χ is the heat of adsorption of a solitary molecule. Thus the true heat of adsorption at $\theta = 0$ is given by

$$q_0 = \chi.$$

When the last molecule evaporates from the surface the change in the potential energy of the system is χ so that the apparent heat of evaporation at $\theta = 0$ is also χ. Thus

$$q_0 = (q_a)_0 = \chi. \tag{5.25 b}$$

The heat of adsorption at $\theta = 1$ can be obtained in the following way. From equation (2·24) it is clear that, as $\epsilon_1 \to \infty$, $\theta \to 1$; from equation (2·21) it is clear that as $\theta \to 1$, $\epsilon_0/\epsilon_1 \to \infty$, so that $\epsilon_0 \to \infty$ as $\theta \to 1$ much more rapidly than does ϵ_1. From equation (2·24) we have

$$\epsilon_1 \sim \frac{3}{4\eta} \frac{1}{1 - \theta} \quad (1 - \theta \text{ small}), \tag{5.28}$$

and so

$$\frac{\partial \epsilon_1}{\partial \theta} \sim \frac{4\eta}{3} \epsilon_1^2 \quad (1 - \theta \text{ small}). \tag{5.29}$$

It follows therefore that, as $\theta \to 1$, the second term of the right-hand member of equation (2·28) approaches minus unity while, for a square lattice, the first term approaches -0.75. Thus, as $\theta \to 1$, $q - q_0 \to -7V$, i.e. the heat of adsorption of the last particle is given by

$$(q_2)_1 = \chi_2 - 7V. \tag{5.30}$$

Since in a completely occupied adsorbed layer each particle which occupies a pair of closest neighbour sites has six interactions with its closest neighbours, it might appear that the heat of adsorption of the last particle should be $\chi_2 - 6V$. For example, consider the sites shown in fig. 26 (a); if all the sites except the pair of closest neighbours marked i and j are occupied, then, when a molecule condenses on i–j, six additional interactions are introduced. Consideration of fig. 26 (a, b), however, shows that if all but two sites are occupied then, in general, the two vacant sites will not be closest neighbours. For, if they are closest neighbours, the interaction energy of the film is $\{\frac{1}{2}(z - 1)N_s - (z + 2)\}V$ (fig. 26 (a)), whereas if they are not closest neighbours the interaction energy of the film is $\{\frac{1}{2}(z - 1)N_s - (z + 3)\}V$ (fig. 26 (b)). Thus, when all but two of the sites are occupied the system will spend most of its time in a configuration such as (b). It follows that, in order to adsorb the last molecule, the particles which are already adsorbed must first rearrange themselves so that they pass from a configuration such as (b) to one such as (a), and this increases the interaction energy by V. The subsequent adsorption of a molecule on a pair of adjacent sites then increases the interaction energy by an additional amount $6V$, so that the heat of adsorption of the last particle is $\chi - 7V$. Thus the true heat of adsorption at $\theta = 1$ is

$$q_1 = \chi - 7V. \tag{5.26 b}$$

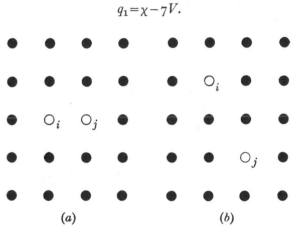

(a)   (b)

Fig. 26. Configurations of occupied sites when only two sites remain vacant.

The apparent heat of evaporation is obtained by considering the evaporation of a molecule from the completely filled surface. No rearrangement of molecules is necessary, and the preceding discussion shows that

$$(q_a)_1 = \chi - 6V. \tag{5.27 b}$$

For adsorption of dimer molecules with dissociation we therefore have, from equations (5.25 a), (5.26 a) and (5.27 a), that

$$(q_2)_1 - (q_2)_0 = -8V, \quad (q_a)_1 - (q_a)_0 = -7V, \tag{5.31}$$

while for adsorption of dimer molecules without dissociation we have, from equations (5.25 b), (5.26 b) and (5.27 b), that

$$q_1 - q_0 = -7V, \quad (q_a)_1 - (q_a)_0 = -6V. \tag{5.32}$$

These elementary considerations show that in each case there is a difference between the true heat of adsorption and the apparent heat of adsorption calculated from the rate of evaporation.

5·6. The Adsorption Isotherm

A great advance in the theory of adsorption was made when Fowler (1935) showed that the Langmuir adsorption isotherm can be deduced from statistical considerations only, without postulating any particular mechanism by which the equilibrium state is attained. In addition, he showed that, when diatomic molecules consisting of two similar atoms dissociate on adsorption and each atom occupies one site, the statistical method leads to an isotherm of the same form as that deduced from kinetic considerations, equation (5·21), on the assumption that there is no interaction between the adsorbed atoms.

One way in which the statistical method is important is that it shows which of the assumptions leading to any particular result are essential and which are irrelevant. For example, in deducing the Langmuir isotherm the essential assumptions are: (i) that there is a definite number of sites for adsorption, (ii) that one gas molecule and one only can be adsorbed on each site, and (iii) that there is no interaction between adsorbed molecules, i.e. that the energies of the states of any adsorbed molecule are independent of the state of occupation of neighbouring sites. The fact that a particular mechanism of condensation and evaporation leads to the correct form of the isotherm merely means that the mechanism has been formulated and written down in a consistent manner. If, in connexion with any particular model of adsorption, we are interested in the mechanisms of condensation and evaporation, it is obviously essential that our formulation of these processes should be consistent with the laws of thermodynamics; that is, it is essential that our formulation of the processes should lead to an isotherm of the same form as that obtained for the same model from statistical considerations.

In order to make the comparison in the present case we must express the constant B in equation (5·17) as a function of the temperature. Let χ_1 be the heat of desorption of an atom from a site surrounded by vacant sites and from the state of lowest energy in the adsorbed phase to the state of lowest energy in the gas phase. Let χ_2 be the heat of desorption, as a molecule, of two atoms adsorbed on neighbouring sites, the surrounding sites being vacant and each atom in its state of lowest energy, the molecule after desorption being in the state of lowest energy in the gas phase.* Let χ_d be the heat of dissociation of a gas molecule in the state of lowest energy into two atoms also in the state of lowest energy. The law of conservation of energy gives

$$\chi_2 = 2\chi_1 - V - \chi_d. \tag{5.33}$$

We now make use of the considerations discussed in § 5·3 and write

$$B_1 = A_1 e^{-\chi_1/kT}, \tag{5.34}$$

$$B_2 = A_2 e^{-\chi_2/kT}, \tag{5.35}$$

and, using the result given in equation (5·11) and neglecting, as discussed in the last footnote, the difference between q_d and χ_d and between $(q_1)_0$ and χ_1,

$$B_4 = A_4 e^{(\chi_1 - \chi_d)/kT}. \tag{5.36}$$

A_1, A_2 and A_4 are to a first approximation independent of the temperature. Further, the van't Hoff equation of the isochore of reaction for the dissociation of hydrogen in the gas phase may be written

$$\frac{\partial \log K}{\partial T} = \frac{\chi_d}{kT^2}.$$

To the present approximation χ_d is treated as a constant. Integrating this we have

$$K = \kappa\, e^{-\chi_d/kT}, \tag{5.37}$$

where κ is independent of T. Using equations (5·33)–(5·37) in (5·14)–(5·16), the latter are all seen to be of the form

$$\frac{\theta}{1-\theta} = A\, e^{(\chi_1 - \frac{1}{2}\chi_d)/kT} p_2^{\frac{1}{2}} \left(\frac{1+\eta\epsilon}{1+\epsilon}\right)^z. \tag{5.38}$$

The expressions for A are respectively

$$\frac{a_1}{A_1}\frac{10^3 K^{\frac{1}{2}}}{\mu_1}, \quad \left(\frac{a_2}{A_2\mu_2}\right)^{\frac{1}{2}}, \quad \frac{A_4}{B_3}\frac{(m_1/m_2)^{\frac{1}{2}}}{10^3 K^{\frac{1}{2}}}.$$

As far as the important terms in T are concerned, equation (5·38) is of the same form as the isotherm which Wang (1937), in his equations (2) and (9), deduced from statistical considerations using the method of Peierls. This agreement is important, for, as has

---

* This differs from the definition of χ_2 given before in that we now specify the change as from one state of lowest energy to another. For practical purposes the difference is negligible.

already been pointed out at the beginning of this section, it means that the formulation of the rates of the various processes is consistent.

We need not illustrate the actual form of the isotherms (see Wang (1937), as at present there are no experimental results available for comparison. Some points of importance in connexion with them may, however, be noted. Wang has shown that, if there is a repulsive interaction between adsorbed particles, the first approximation treatment which we have given leads to an isotherm in which there is no point of inflexion. Chang (1938) has shown that, if the approximation is carried further and the effect of long-distance order is included, there is in some cases a point of inflexion, but numerical calculation shows that this occurs only when θ varies slowly with p, and it would be difficult to detect the effect experimentally. In connexion with points of inflexion in isotherms it has often been stated that, if the surface is not uniform in that various sites on it have different energies of adsorption associated with them, quite apart from any interaction between the adsorbed particles, the adsorption isotherm will show steps. Wang (1937, p. 136) has shown that this is not so.

CHAPTER 6

SOME OTHER TYPES OF ADSORPTION

6·1. The Distribution of Particles in a Mobile Film when each Adsorbed Particle precludes Occupation of Neighbouring Sites

In applying the theory of adsorption developed in the preceding chapters to the adsorption of hydrogen, it has been assumed that each hydrogen molecule dissociates on adsorption, and each adsorbed atom occupies one site which has four closest neighbour sites. It was pointed out in § 1·4 that the amount of hydrogen adsorbed on tungsten could equally well be explained on the view that the hydrogen is adsorbed as molecules without dissociation, and that, if a given site is occupied, the size of the molecules is such that no adsorption can occur on the four neighbouring sites. This consideration also applies in examining the amount of oxygen in the stable film on tungsten. When such a film is complete there are half as many adsorbed molecules as there are metal atoms in the surface, i.e. the number of adsorbed gas atoms is equal to the number of metal atoms in the surface. It is therefore important to develop the theory of this type of film to see if there is any crucial experimental test which would enable us to decide definitely to which of the two types a given film belongs.

The theory of the production of atomic hydrogen on the basis of adsorption with dissociation accounts for the experimental results satisfactorily (§ 5·4). This suggests that at the temperatures 1150–1420° K. at which the experiments were carried out the hydrogen film is an atomic one. Blodgett and Langmuir (1932), in the experiments already mentioned in § 1·2, measured the accommodation coefficient a of hydrogen with a tungsten filament at a mean temperature of 357° K., the hydrogen being at 80° K. They showed that, if the filament was heated in a hydrogen atmosphere to a temperature of about 600–1000° K. before the measurement was made, the value of a obtained was lower than that given by a filament that had not been heated higher than about 400 or 500° K. after admitting the hydrogen. They interpreted this result as meaning that there are two types of adsorption of hydrogen on tungsten, and suggested that the simplest assumption was that at low temperatures the hydrogen is adsorbed in the molecular form, while in a film formed at higher temperatures it is atomic. They point out that Bonhoeffer and Farkas (1931) have suggested that the *ortho-para*-hydrogen conversion of hydrogen on tungsten can be explained if it is assumed that a film formed at room temperature is atomic, and that, if this contention is accepted as correct, it would appear to be necessary to conclude from their experiments that there are two types of atomic adsorption.

There is a real difficulty here because, whatever view may be taken of the detailed mechanism by which the conversion goes, it is difficult to explain its occurrence on the surface without postulating dissociation of the hydrogen when it is adsorbed at room temperature. There is the further consideration in favour of dissociation on adsorption that it is difficult to picture how the hydrogen can be held by chemical forces unless it is dissociated into atoms, because the electrons in the molecule form a closed shell. On the other hand, if the atomic nature of the film at room temperature is accepted, it seems equally difficult to devise a plausible model of the two types of atomic adsorption demanded by Blodgett and Langmuir's result. The weight of the evidence appears to be in favour of an atomic film to which the considerations brought forward in this chapter will not apply. With oxygen, which we now consider, these difficulties do not arise.

Johnson and Henson (1938) have used the method developed by Johnson and Vick (1935) (§ 5·5) to study the formation of an oxygen film on tungsten at high temperatures (2100–2400° K.). They have found that in this temperature range there is an apparent energy of activation for the adsorption process. The experiments indicated that under given conditions the rate of adsorption increases with the temperature and, if the same law continued to ordinary temperatures, it would therefore be negligibly small. They suggest, therefore, that in this case there are two types of adsorption, some sort of molecular adsorption at low temperatures and atomic adsorption at high, the occurrence of the latter requiring an energy of activation. Here, then, there is sufficient evidence to suggest the desirability of examining the properties of molecular films.

We have already seen in § 1·4 that there is some doubt about the actual crystal plane that predominates in the surface of a tungsten filament. From the present point of view this is of considerable importance, as the 110 and the 100 planes would behave differently. We shall first consider the 110 plane in which the atoms are arranged as shown in fig. 27, and shall assume that, if a molecule is adsorbed on a given site (say site 0 in the figure),

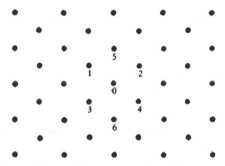

Fig. 27. Group of sites on 110 plane of tungsten.

the neighbouring sites (1, 2, 3 and 4) are not available for the adsorption of another molecule. We assume that, when two molecules are adsorbed on sites like 0 and 5, the nearest sites on which adsorption can take place, there is an interaction energy V between them due to their mutual repulsive forces, but that at greater distances, for example on sites 1 and 2, the interaction energy is negligible.

Let us first consider the states of lowest energy for a film of this type. Up to $\theta = \frac{1}{3}$ it is possible, as shown in fig. 28, to arrange the particles on the surface so that the interaction energy is zero. The small dots in the figure represent vacant sites and the larger ones particles. It is not possible to put more particles on the surface or to arrange them differently without introducing interactions. Up to $\theta = \frac{1}{3}$ the heat of adsorption will remain constant at q_0. An examination of fig. 28 shows that no further adsorption can occur without breaking up this configuration. Some rearrangement such as the following is necessary. Particle 1 moves to site 3 and particle 2 to site 4 simultaneously. Particle 5 then moves to site 6 and site 7 is available for the adsorption of an extra particle. In order to put in this one extra particle we have introduced six interactions and its heat of adsorption is anomalous. The extra energy can be regarded as a sort of activation energy for starting the rearrangement right along the chain AB, since, once the rearrangement has proceeded to this extent, particle 8 can move to site 9 leaving site 10 vacant for an extra particle and so on along the line on both sides, each particle added introducing three interactions, so that the heat of adsorption per particle is $q_0 - 3V$. This holds till

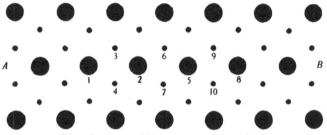

Fig. 28. State of lowest energy when $\theta = \frac{1}{3}$.

we get to the ends of the line where actually the heat is higher than $q_0 - 3V$. When the total number of sites is large, the anomalies at the beginning and end of a line are negligible, and the relation between $(q - q_0)/V$ and θ is shown by the heavy line in fig. 29. This behaviour will be approached when V/kT is large, i.e. when $\eta = e^{-V/kT}$ approaches zero, and the curve is marked $\eta = 0$. When, on the other hand, V/kT is very small, i.e. $\eta \sim 1$, the distribution of particles on the surface is random at all stages of filling up the film, and there is never any necessity to break up a configuration of lowest energy. The heat of adsorption for $\theta = 0.5$ is the heat corresponding to the filling up of the last gap in this random distribution and is $q_0 - 2V$. For the random distribution the relation between heat of adsorption and θ is approximately linear* and is

* The relation is not strictly linear because, even when there is no appreciable interaction between adsorbed particles on sites like 0 and 5 in fig. 27, i.e. V/kT small, the fact that site 0 is a site on which adsorption can take place affects the state of occupation of site 5. In other words, site 5 is not an average site, the reason for this being that, if adsorption can take place on the vacant site 0, sites 1 and 2 must also be vacant, and a site which is a neighbour to two vacant sites is not an average site. Only for an average site can it be stated that the probability of its being occupied is θ. Thus, when site 0 and its four neighbours are vacant, the probability of finding site 5 occupied is not exactly θ. The relation between heat of adsorption and θ would be strictly linear only if this probability were θ.

shown by the curve marked $\eta = 1$ in fig. 29. The actual shape of this curve and the shapes of those corresponding to values of η between 0 and 1 have been calculated by Roberts (1938b).

The behaviour in the region $\theta = 0.5$, when the film is nearly complete, is remarkable. For a given system, i.e. a given value of V, the heat of adsorption varies markedly with T and increases as T increases. It is important to note that for this type of system this does not, as might be supposed, *necessarily* mean that, as the temperature is raised, the type of adsorption changes. Any further discussion of this point would be premature as yet, but it must be

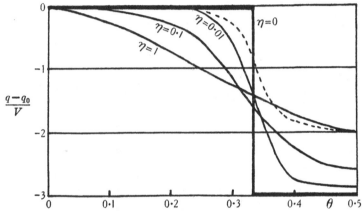

Fig. 29. Variation of the heat of adsorption with the fraction of the surface covered. The full lines are true heats of adsorption; the broken line gives apparent heat of adsorption calculated for $\eta = 0.01$.

pointed out that very complicated variations of heat of adsorption with temperature and with θ can arise for quite simple systems when there are forces between the adsorbed particles.

6·2. The Kinetics of Adsorption and Evaporation in a Mobile Film on the 110 Plane of Tungsten when each Adsorbed Particle precludes Occupation of Neighbouring Sites

Similar considerations will apply to the rate of condensation during the formation of the film, and for small values of η we should expect a rapid diminution in this rate in the neighbourhood of $\theta = \frac{1}{3}$.

Let n_s be the number of sites per unit area. If re-evaporation of particles is negligible, $n_s d\theta/dt$, the rate of condensation per unit area is given by

$$n_s \frac{d\theta}{dt} = \frac{\alpha p}{\mu} \phi(\theta), \tag{6.1}$$

where p is the gas pressure in dynes cm.$^{-2}$, $\mu = (2\pi mkT)^{\frac{1}{2}}$ and α is the condensation coefficient, i.e. the chance that a molecule condenses when it strikes a vacant site surrounded by four vacant

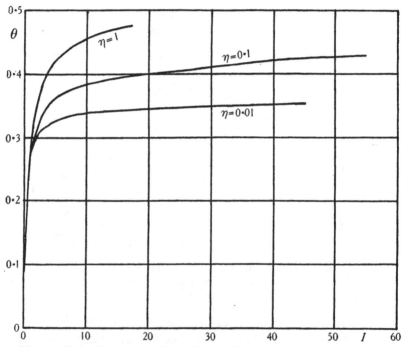

Fig. 30. Rate of formation of film. The abscissae, which are the values of the integral on the right-hand side of equation (6.2), give the time in arbitrary units.

sites. $\phi(\theta)$ is the probability that any given site is vacant and is surrounded by four vacant sites. Roberts (1938 b) gives $\phi(\theta)$ as a function of θ and the appropriate ϵ parameters, introduced in the Bethe approximation, which are given implicitly as functions of θ by Roberts's equations (5) and (6). If we integrate this equation, it becomes

$$\frac{\alpha p}{\mu n_s} t = \int_0^\theta \frac{d\theta}{\phi(\theta)} = I, \tag{6.2}$$

where $\theta = 0$ when $t = 0$. Values of this integral have been calculated by Roberts (1938b) (equation (12)) for various values of η. Since I is proportional to t we can see how θ varies with t by plotting the relation between θ and I. This is done in fig. 30, and it will be seen that for small η there is, as we should expect, a rapid diminution in the rate of formation of the film near $\theta = \frac{1}{3}$.

A formula for the rate of evaporation r per unit area can also be obtained statistically, and the apparent heat of evaporation q_a can be deduced from the formula

$$q_a = -k \frac{d \log r}{d(\mathbf{1}/T)}.$$

Values of $(q_a - q_0)/V$ obtained in this way for $\eta = 0.01$ are shown by the broken line in fig. 29. It will be seen that in this case, as in that considered in § 5·5, there is a marked difference between the apparent heat deduced in this way and the true heat. Even for this small value of η the value of $(q_a - q_0)/V$ is -2 at $\theta = 0.5$. The reason for this is evident when the process of evaporation from a complete film is considered. The heat term in the exponential part of the expression for the rate of evaporation is simply the energy required to remove a single particle from the complete film. This is independent of η and is the apparent heat obtained from the application of the formula which is given above.

6·3. Properties of Immobile Films on Tungsten in which each Adsorbed Particle precludes Occupation of Neighbouring Sites

If the particles in the adsorbed film are immobile in the sense discussed in § 2·2 and there is no re-evaporation, and if the probability that a molecule striking a suitable vacant site on the surface will condense is independent of the state of occupation of near sites, the relation between heat of adsorption and fraction of sites occupied is approximately linear.

Whenever the spacing of the adsorbed particles in an immobile film differs from that of the underlying solid atoms, Roberts (1935 b) has shown that the final film will have gaps in it. In the present case Roberts (1938 b) obtained the number of gaps by using an empirical method which involved the selection of random sites on a model of the surface. If there are N_s sites for adsorption in all, and if for a given value of θ there are N sites available for adsorption, we define $f(\theta) = N/N_s$. If the film were arranged in such a way that it contained the maximum number of particles, $f(\theta)$ would be zero when $\theta = 0.5$. Actually it was found to be zero when $\theta = 0.4$, so that the effect of the gaps is to diminish the amount in the final film by about 20 %.

An estimate of the fraction of gaps in this case can also be found by a method similar to that which was used in § 4·1 for the case in which each particle occupies two sites on the surface. In the present case a vacant site is not available for adsorption of a particle if any of its closest neighbour sites are occupied. The probability that this will be so can be found* as a function of the number of sites occupied. This function has a maximum value of 0·18 in the neighbourhood of $\theta = 0.21$. This is a lower limit to the number of gaps in the complete immobile film because as further particles are adsorbed on the sites still available, the nearest neighbours of additional vacant sites will become occupied. Thus, this estimate is in agreement with that deduced empirically.

6·4. General Summary

In making a general comparison between the different types of film we shall first summarize the differences between films formed on the 110 plane of tungsten (a) by the dissociation of diatomic molecules into two similar atoms, each atom occupying one site, and (b) by adsorption without dissociation, the size of the molecules being such that, if a molecule is adsorbed on a given site, the four neighbouring sites are not available for the adsorption of another molecule. We assume that there is a large repulsive interaction between adsorbed particles, i.e. that V/kT is fairly large.

Let us first consider mobile films in equilibrium. For the two types of film the qualitative behaviour will be similar, and at a certain stage in the process under consideration there will be rapid changes in the heat of adsorption or the apparent heat of evaporation and in the rate of formation of the film if there is no appreciable re-evaporation. For the film formed with dissociation the changes occur when it contains one-half the maximum number of

* Miller, A. R., unpublished results.

particles and for the other type when it contains two-thirds the maximum number.

With immobile films the relation between the heat of adsorption and the amount adsorbed is nearly linear in both cases. The variation in the rate of formation with the amount adsorbed is not very different in the two cases, but for the film formed with dissociation the amount in the final film differs by 8 % from that in a complete mobile film, while for the other type the amount in the final film differs by about 20 % from the amount in the complete mobile film. To show that any difference between the variation in the rate of formation with amount adsorbed does not provide an experimental method for distinguishing between the two types the values of $\int_0^\theta \dfrac{d\theta}{\phi(\theta)}$ in equation (6·2) have been plotted as dots in fig. 21, using Bosworth's experimental data to determine the relation between θ and t. These dots refer to a film in which the particles are adsorbed as molecules and are so large that they preclude adsorption on neighbouring sites. As has already been seen in § 4·1 the crosses are for a film formed by dissociation and the adsorption of the two atoms on single sites.

If, on the other hand, the 100 plane predominates, the surface atoms are arranged in a simple square lattice. For a mobile film on a lattice of this type the changes in the rate of formation and in the heat of adsorption occur when the film is half complete for the film formed with dissociation. The change in the heat also occurs when the film is half complete for molecular adsorption in which the final film contains half as many molecules as there are atoms in the surface. There is no effect of interaction energy on the rate of formation of the molecular film. Thus the two types of mobile film could be distinguished by studying the rate of formation. For an immobile film formed with dissociation the final amount would differ by 8 % from that in a complete mobile film, while for an immobile film formed without dissociation the final amount would differ by about 20 % from that in a complete mobile film, also when adsorption takes place on the 100 plane.

The behaviour is summarized further in Table 5. It will be seen from the table that a definite decision between the two types of adsorption can be obtained, whichever plane predominates, by a study of the difference between the amounts in the final immobile film and the complete mobile film. The results obtained by Van Cleave with oxygen on tungsten and given in fig. 22 have been tentatively interpreted as being due to the difference between the amounts in the final immobile film and in a complete mobile film. The accommodation coefficient for the final immobile film is 0·2 and it rises to 0·24 when the film is heated, this rise being interpreted as due to the possibility of adsorbing additional oxygen in a mobile film. The accommodation coefficient for a bare surface is 0·06 and, if we assume that the change in accommodation coefficient is proportional to θ, we have for the ratio of the amount in the final immobile film to that in the complete mobile film 0·14/0·18 = 0·78. This agrees with 0·8, which would be expected for the molecular film, but not with 0·92, which would be expected for the film formed with dissociation. In so far, then, as the interpretation of the nature of the rise in accommodation coefficient and shown in fig. 22 is correct, this suggests that the oxygen film is molecular.

TABLE 5

Value for mobile film of θ/θ_f at which change in true and apparent heats and in rate of formation occurs				Ratio of amounts in final immobile film and in complete mobile film			
100 plane		110 plane		100 plane		110 plane	
Dissociation	Without dissociation *	Dissociation	Without dissociation	Dissociation	Without dissociation	Dissociation	Without dissociation
0·5	0·5	0·5	0·67	0·92	0·80	0·92	0·80

θ_f = value of θ when film is complete.

* There is no effect on the rate of formation in this case due to the interaction energy.

Some experiments on the properties of oxygen films on tungsten by Morrison and Roberts (1939) have a bearing on this question and illustrate the general methods that have been developed. We now turn to a brief account of these experiments.

6·5. Experimental Method

In the experiments of Morrison and Roberts (1939), the accommodation coefficient method (§ 1·2) was modified in such a way as to enable the actual pressure of oxygen at the wire to be obtained. The principle of the method will be understood by reference to fig. 4, although the actual apparatus differed in detail from that illustrated there. The oxygen was admitted from a gas pipette to a large calibrated bulb of about 2 litres capacity, so that the partial pressure in this bulb was known. It then passed through a long fine capillary tube and entered the apparatus at a point near the side tube leading to the Macleod gauge. Here it divided into three streams. The first filled the Macleod gauge and, once equilibrium was established, ceased. The second flowed to the near charcoal tube and was taken up by the charcoal. The third flowed past the wire to the other charcoal tube. The resistance to the flow of oxygen arose in part from the effect of collisions with the walls of the tubes and in part from the fact that it was diffusing through neon. The dimensions of the tubes were so chosen that in the capillary tube the resistance was primarily due to collisions with the walls, the effect due to the presence of the neon being only a correction, while in the other parts of the apparatus the resistance was primarily due to the presence of the neon and the wall effect was only a correction. Under these conditions the flow could be calculated accurately, and the partial pressure of the oxygen at any point, and in particular at the wire, could be deduced. The conditions obtaining at the surface of the charcoal were investigated in subsidiary experiments, and it was found that a small correction was necessary for the fact that the charcoal was not perfectly efficient in taking up oxygen. The pressure of oxygen at the wire depended markedly on the neon pressure, and it was a satisfactory test of the theory of the method to find that consistent results were obtained when this pressure was varied in the ratio 2 to 1.

6·6. Three different Types of Oxygen Film

With the wire used the accommodation coefficient of neon with bare tungsten was 0·057. When oxygen was admitted and adsorbed on the tungsten, this rose to a final steady value a_f, which depended on the pressure of oxygen at the wire. a_f is plotted against the oxygen pressure in fig. 31. Extrapolating the curve, we see that at very low pressures the value of $a_f = 0.226$ would be reached, although of course the time taken to reach it would be considerable. This suggests two types of adsorption: a very stable film corresponding to $a = 0.226$ and a less stable one on top of it, the population in which depends on the pressure. Further investigation showed, however, that the film corresponding to $a = 0.226$ is itself composite. Heating this film to a temperature of about 1100° K. reduced the accommodation coefficient to 0.177, and, until it was heated to above 1700° K., no further evaporation took place.*

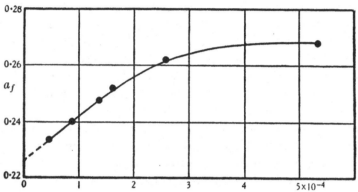

Fig. 31. Relation between the final steady value of the accommodation coefficient of neon (a_f) and the partial pressure of oxygen.

The stable film which does not begin to evaporate appreciably until above 1700° K. can be presumed to be immobile, that is, each particle remains where it is first attached. If it is assumed that the film is atomic, each molecule dissociating on adsorption and the two atoms occupying two neighbouring vacant sites, the final film will have 8% of holes in it consisting of isolated uncovered single sites (see § 4·1). As we have seen in § 4·2, it would be expected that such sites would exert a greater attraction for impinging oxygen molecules than the other parts of the surface,

* The value 0·177 corresponds to 0·20 given by Van Cleave (1938) and referred to in § 4·2. These small differences are not relevant from the present point of view and may be due to slight variations in the roughness of different samples of tungsten. The rise in the value of the accommodation coefficient when the tungsten was heated above about 1400° K. shown in fig. 22 was also observed in the present experiments.

and that oxygen molecules would be strongly adsorbed on them. The most likely explanation of the results is then that the state $a = 0.177$ produced by heating the film to about 1100° K. corresponds to the stable atomic film with 8 % of gaps in it unfilled. The state $a = 0.226$ reached at very low pressures corresponds to the stable film with the 8 % of gaps occupied by molecules. The higher values, shown in fig. 31, are due to the presence of an adsorbed film on top of this structure, the equilibrium amount in this last film depending on the pressure of oxygen in the gas phase.

The form of the adsorption isotherm for this last film was investigated. The highest pressure at the wire which could be measured accurately was about 9.8×10^{-3} dyne cm.$^{-2}$. At higher pressures than this the necessary conditions for the theory to be applicable were not satisfied. At 9.8×10^{-3} dyne cm.$^{-2}$ the final steady value a_f of the accommodation coefficient was 0.320. When the pressure of the oxygen in the large bulb from which the oxygen diffused through the capillary tube was increased 4 times, the value of a_f was 0.340, and, when it was increased about 10 times, a_f was 0.352. The pressures at the wire in these two experiments could not be deduced accurately, but they must have been of the order of 4 and 10 times that in the experiment in which a_f was 0.320. It is thus evident that at these pressures the value of a_f is changing very slowly with increase of pressure, and that the final value corresponding to a complete film cannot lie far from 0.36.

Let us consider the 100 plane of tungsten (see fig. 8). The sites in the upper layer we are considering will be arranged in the same way as the atoms in the surface, so that the distance between nearest neighbours will be 3.15×10^{-8} cm. The diameter of the oxygen molecule is (Jeans 1925) 3.64×10^{-8} cm. Thus it is probably not possible to pack oxygen molecules on the surface so that every site is occupied, but easily possible to pack them so that, if a given site is occupied, the surrounding four sites are vacant as considered in § 6·1 for the 110 plane. In this case the complete film with $a_f = 0.36$ will correspond to $\theta = 0.5$, if we neglect the fact that the underlying structure is not completely uniform, in that places where undissociated molecules are adsorbed on vacant sites surrounded by four occupied sites will differ from sites occupied by atoms. We therefore take the value of θ corresponding to any value of a_f as

$$\theta = 0.5 \frac{a_f - 0.226}{0.36 - 0.226}. \tag{6.3}$$

In this way the experimentally realised adsorption isotherm can be determined by using equation (6.3) to calculate the fraction of the surface covered corresponding to the pressure at which a_f was measured. This isotherm can be compared with that determined theoretically, by using Bethe's approximation to examine* the adsorption of molecules on a regular square array, as the 100 plane of tungsten, to form an immobile film when the occupation of any site on the surface precludes the occupation of its four closest neighbours by other molecules. In the first approximation one finds that ξ, which is proportional to the pressure, is given as a function of θ by

$$\xi = \frac{\theta(1 - \theta)^3}{(1 - 2\theta)^4}. \tag{6.4}$$

This gives very good agreement with the experimentally determined isotherm.†

From the adsorption isotherm, in the range of low values of θ, where the relation between the gas pressure and the fraction of the surface covered is almost linear, it can be shown that the mean life, τ, of an adsorbed molecule in this second layer is not less than about 2 seconds. The probability that in a time Δt ($\Delta t \ll \tau$) a given particle on the surface will evaporate is $\Delta t/\tau$. If n_s is the number of sites per unit area of a smooth surface and ρ is a factor to take account of the roughness then there are $\rho\theta n_s$ molecules adsorbed per unit apparent area of surface. Then the number of molecules which evaporate in a time Δt is

$$n_e = \rho\theta n_s \, \Delta t/\tau.$$

---

* Miller, unpublished results.

† In considering the adsorption isotherm Morrison and Roberts (1939) quote a formula given earlier (Roberts 1938b), in considering an array similar to the 110 plane of tungsten. This formula is obtained in the second approximation and they state that for an immobile film this formula also applies to adsorption of the kind under consideration on a regular square array. This is not so, owing to the difference in symmetry of the two planes. In any case, owing to the size of the oxygen molecule and the lattice spacing, the results of the analysis for the earlier model (Roberts 1938b) could not be used to discuss the adsorption of oxygen on tungsten.

For small values of θ, the number of molecules which condense on unit apparent area of surface in time Δt is

$$n_c = a \frac{p}{(2\pi mkT)^{\frac{1}{2}}} \Delta t,$$

where a is the condensation coefficient. In equilibrium we therefore have

$$\theta = \frac{a}{\rho} \frac{\tau}{n_s} \frac{p}{(2\pi mkT)^{\frac{1}{2}}}. \tag{6.5}$$

For the 100 plane of tungsten, $n_s = 10^{15}$ so, at room temperature,

$$\theta = 2.67 \times 10^2 (a/\rho)\, \tau p.$$

In the region in which there is a linear relation between the accommodation coefficient and the gas pressure it can be seen from fig. 31 that when $p = 10^{-4}$ dyne cm.$^{-2}$, $a_f = 0.242$ so that $\theta = 0.06$. These values yield

$$\frac{a}{\rho} \tau = 2.25.$$

Thus, since a/ρ cannot be greater than unity, the average life of an adsorbed molecule in the second layer is not less than a little over 2 sec.

It is important to consider the actual changes in a. During the formation of the first stable layer with the gaps unfilled, a changes by 0.120. When the gaps are filled, a changes by 0.049. When the second layer is formed, a changes by 0.134. If there are only 8 % of gaps, the second change seems disproportionately large compared with the other two. If, on the other hand, the oxygen in the first stable layer were adsorbed as undissociated molecules, the adsorption of such a molecule on a given site precluding the possibility of adsorption on the four neighbouring sites, as considered in § 6.3, the gaps in the film reduce the amount in it by 20 %. It has been shown (Roberts 1940) that the number of molecules that can be adsorbed per unit area over these gaps is in the neighbourhood of $0.2 n_s$, while the number in the second layer is in the neighbourhood of $0.5 n_s$. Thus for a film of this type it would be much easier to account for the relative changes in a in the different stages. We have already seen in § 6.4 that another result for oxygen films adsorbed on tungsten is in more satisfactory numerical agreement with what would be expected for this type of film than with what would be expected for a film in which the first stable layer was atomic.

Whatever may be the nature of the first stable film, the essential point that is established by these experiments is that there are the three types of adsorbed film, and that the most reasonable way of accounting for these is to suppose that the first consists of a very stable film on the tungsten, that the second consists of molecules adsorbed in the gaps in this stable film, and that the third consists of molecules in a second layer.

6·7. The Kinetics of the Adsorption of Oxygen

It is easy to see that the kinetics of the formation of the oxygen film will be profoundly affected by the existence of the upper layer. A molecule which strikes a site in the stable film which is already occupied is not immediately permanently adsorbed, but the fact that there is a second molecular layer formed, the amount in which depends on the pressure of the gas, must mean that molecules have a finite life on the surface in this upper layer, where they will presumably be mobile. Thus, even if the first site which a molecule tries is occupied, it may still find its way to a vacant place in the stable layer by moving over the surface. Let θ be the fraction of sites occupied in the stable layer. The effect of the presence of the upper mobile layer is to make the rate of adsorption into the stable layer vary slowly with θ, particularly when θ is small.

It has been seen that the behaviour of the oxygen film on tungsten is consistent with the view that the most stable film is atomic, forming an immobile film. In such a film 8 % of the total number of sites are vacant, that is, this fraction of single sites is surrounded by occupied sites (§ 4·1). In the oxygen films these sites can be occupied by molecules, and this completes the first adsorbed monolayer. Let there be n_s sites per unit area of a smooth surface. Let θn_s be the number of sites per unit area that are occupied by the stable atomic film and let ϕn_s be the number which are occupied by the second kind of film, that is, by molecules on single sites surrounded by four sites occupied by other molecules. The chance that any given site is occupied is $\theta + \phi$, so that if a molecule striking the surface evaporates before it can migrate to another site in the second layer, the chance that it is reflected is $\theta + \phi$. The effect of the mobility of the particles in the second layer can be allowed for by supposing that when the first layer is completely filled, that is, $\theta + \phi = 1$, each molecule which strikes the surface migrates, *on the average*, to r sites (additional to the site at which it first strikes the surface) before it evaporates. When the first adsorbed layer on the surface is only partly filled the probability that all the $r + 1$ sites at which the molecule impinges on the surface are occupied is

$$(\theta + \phi)^{r+1}.$$

This gives the probability that any molecule striking the surface evaporates. The fraction of molecules striking the surface which are adsorbed is

$$1 - (\theta + \phi)^{r+1}, \tag{6.6}$$

provided the states of occupation of all the sites which a molecule tries are independent. This assumption cannot be strictly correct but, provided the actual population of the second layer is extremely small, the neglect of any mutual effects of molecules in the second layer can be neglected, and the results obtained will be valid to a first approximation.

For an immobile film formed by the adsorption of the atoms of a diatomic molecule on closest neighbour sites it has been seen (Roberts and Miller 1939) that if a given site is vacant, the probability that the four sites surrounding it are all occupied is

$$\left(\frac{\epsilon_1}{1 + \epsilon_1}\right)^4, \tag{6.7}$$

where ϵ_1 is determined, as a function of θ, by

$$\epsilon_1 = \frac{3}{4} \frac{\theta}{1 - \theta}. \tag{6.8}$$

In order that a site may be suitable for adsorption into the first or atomic layer it is necessary that it should be unoccupied and that at least one of its closest neighbours should also be unoccupied, that is, that it should not be surrounded by four occupied sites. The probability of the first of these conditions is $1 - \theta$ and of the second is, from equation (6.7),

$$1 - \left(\frac{\epsilon_1}{1 + \epsilon_1}\right)^4.$$

Since we are concerned with the random occupation of the array of sites, it follows that the probability that a given site is suitable for adsorption of a molecule into the first kind of film (stable atomic film) is

$$(1-\theta)\left\{1-\left(\frac{\epsilon_1}{1+\epsilon_1}\right)^4\right\}. \tag{6.9}$$

Likewise, the probability that any given site is occupied neither by an atom nor by a molecule (forming part of the second kind of film) is

$$(1-\theta)\left(\frac{\epsilon_1}{1+\epsilon_1}\right)^4-\phi. \tag{6.10}$$

Assuming that the state of occupation of any site is independent of that of any other site, we then get, from the expressions (6.8), (6.9) and (6.10), that

$$\frac{d\theta}{dt}=\frac{2}{\rho n_s}\frac{p}{(2\pi mkT)^{\frac{1}{2}}}(1-\theta)\left\{1-\left(\frac{\epsilon_1}{1+\epsilon_1}\right)^4\right\}\frac{1-(\theta+\phi)^{r+1}}{1-(\theta+\phi)}, \tag{6.11}$$

$$\frac{d\phi}{dt}=\frac{1}{\rho n_s}\frac{p}{(2\pi mkT)^{\frac{1}{2}}}\left\{(1-\theta)\left(\frac{\epsilon_1}{1+\epsilon_1}\right)^4-\phi\right\}\frac{1+(\theta+\phi)^{r+1}}{1-(\theta+\phi)}, \tag{6.12}$$

since $p/(2\pi mkT)^{\frac{1}{2}}$ molecules strike unit apparent area per second. From these equations we get

$$\frac{d\phi}{d\theta}+\phi/2(1-\theta)\left\{1-\left(\frac{\epsilon_1}{1+\epsilon_1}\right)^4\right\}-\left(\frac{\epsilon_1}{1+\epsilon_1}\right)^4\bigg/2\left\{1-\left(\frac{\epsilon_1}{1+\epsilon_1}\right)^4\right\}=0. \tag{6.13}$$

It should be noted that this equation does not involve r. Further, ϵ_1 is given explicitly as a function of θ by equation (6.8) so that since equation (6.13) is a linear equation, it can be integrated. The constant of integration is determined by the fact that θ and ϕ vanish together. This determines ϕ as an explicit function of θ. Thus, using this result in equations (6.11) and (6.12), $d\theta/dt$ and $d\phi/dt$ can be determined for any value of θ.

It has been noted already in § 6.6 that the accommodation coefficient for the bare surface is 0·057 and that for the stable atomic film, for which 92 % of the sites are occupied, the accommodation coefficient is 0·177. An apparent value of θ can thus be determined by

$$\theta_a=\frac{a-0\cdot057}{0\cdot130}.$$

If now we assume that there is a linear relation between the accommodation coefficient and θ and ϕ, we can write

$$a = 0.057 + 0.130\theta + 0.613\phi,$$

where the coefficient of ϕ has been determined by the fact that when the first adsorbed layer is completed, that is, $\theta = 0.92$ and $\phi = 0.08$, the accommodation coefficient is 0.226. Thus

$$\theta_a = \theta + 4.72\phi. \tag{6.14}$$

Using the values of $d\theta/dt$ and $d\phi/dt$ which can be determined from equations (6.8), (6.11), (6.12) and (6.13), equation (6.14) can be used to determine $d\theta_a/dt$ as a function of θ.

Using the appropriate values for the 100 plane of tungsten we get, for the theoretical value of $d\theta_a/dt$,

$$\frac{d\theta_a}{dt} = \frac{d\theta}{dt} + 4.72\frac{d\phi}{dt},$$

where

$$\frac{1}{p}\frac{d\theta}{dt} = \frac{5.34 \times 10^2}{\rho}(1-\theta)\left\{1 - \left(\frac{\epsilon_1}{1+\epsilon_1}\right)^4\right\}\frac{1-(\theta+\phi)^{r+1}}{1-(\theta+\phi)} \tag{6.15}$$

and

$$\frac{1}{p}\frac{d\phi}{dt} = \frac{2.67 \times 10^2}{\rho}(1-\theta)\left\{\left(\frac{\epsilon_1}{(1+\epsilon_1)}\right)^4 - \phi\right\}\frac{1-(\theta+\phi)^{r+1}}{1-(\theta+\phi)},$$

where ϵ_1 and ϕ are determined as functions of θ by equations (6.8) and (6.13).

By working at very low pressures and measuring the gradients of curves giving the relation between the accommodation coefficient and the time, Roberts and Morrison (1939) were able to investigate the variation of $d\theta_a/dt$ with θ. At these pressures the amount actually present in the upper layer at any instant was very small, and any direct effect of it on the accommodation coefficient was negligible. The results obtained were in agreement with what would be expected for the type of system to which oxygen has been shown in § 6.6 to belong. It may be mentioned that, although it is difficult to measure gradients accurately, this is the best way to use the results, because the relation between $d\theta_a/dt$ and θ is much more characteristic and varies much more from one type of film to another than other relations that can be used, for example, the relation between θ and t.

DIPOLE INTERACTIONS BETWEEN ADSORBED PARTICLES

7·1. Introduction

We have seen that the variation of the heat of adsorption with amount adsorbed is very different for different types of film, and that an experimental determination of this variation can give information as to the nature of the film. Up to the present we have considered only those cases in which the force between the adsorbed atoms is repulsive and where it falls off so rapidly with distance that only the interaction between nearest neighbours need be considered. In many cases of the adsorption of gases on solids, long-range forces which are of electrostatic origin and which usually consist of dipole-dipole interactions have to be taken into account. In this chapter we shall calculate the effects that may arise when the interaction between adsorbed molecules arises partly from van der Waals *attractive* forces and partly from repulsions between permanent or induced dipoles, and shall consider the bearing of the calculations on a number of experimental results.

It has, of course, been pointed out in various connexions that permanent or induced dipoles in adsorbed molecules will affect heats of adsorption. For example, if a molecule with a permanent dipole moment is adsorbed on the surface of a metal, there is an attractive force between the dipole and the metal, and the potential energy can be calculated using the method of electric images.* This potential energy will contribute to the heat of adsorption.

If all the ions in the surface plane of an ionic crystal have the same charge there is an electrostatic field near the surface. Lenel (1933) has pointed out that the presence of such a field which will induce dipoles in adsorbed molecules can account for the difference between the heats of adsorption of argon on KCl (simple cubic) and on CsCl (body-centred cubic) crystals.† Lennard-Jones and

---

* See, for example, de Boer (1935).

† Bradley (1936) has worked out a theory of the formation of multilayers based on the view that the induced dipoles in one layer will themselves induce dipoles in the layer above. See also de Boer and Zwikker (1929).

Dent (1928) have considered the electrostatic field near the surface of an ionic crystal in which each ion is surrounded by four ions of different sign (e.g. simple cubic KCl). They have pointed out that there is a strong field immediately above a given ion, but that the energy of adsorption due to electrostatic forces *plus* van der Waals forces is in fact greatest above a point equidistant from four ions where the electrostatic field is zero. Blüh and Stark (1927) have also considered the energy of interaction of a rare gas atom and an alkali halide crystal. These calculations were concerned with the actual heat of adsorption of a solitary particle on the surface.

Let us then consider molecules with permanent dipole moments adsorbed on a surface in such a way that the moments are all parallel to each other and normal to the surface. If the surface is a conductor and there is only one molecule on it, there will be a term in the potential energy arising from the attraction between the dipole in the molecule and its electric image. When the surface is fully covered, the corresponding term will arise from the force between the dipole in the molecule and its own electric image together with the electric images of all the other adsorbed molecules. All the electric images taken together approximate to two infinite parallel sheets of charges of opposite sign, and their net effect is smaller than that of a single image. Thus, as the surface becomes covered, this term in the heat of adsorption undergoes considerable variation in such a way as to make the heat of adsorption decrease. The direct repulsive forces between neighbouring adsorbed dipoles also causes the heat to decrease as the surface becomes covered.

For films of caesium atoms held on the surface of tungsten as positive ions, each ion and its electric image are equivalent to a dipole. In this connexion, Langmuir (1932) has pointed out that a number of parallel dipoles exert mutual depolarizing effects on each other, and he and Taylor (1933) have shown directly that the electric moment per adsorbed caesium atom diminishes considerably as the number of adsorbed atoms increases. We shall not consider films in which the adsorbed atoms are held as positive ions, but those in which the adsorbed molecules have permanent or induced dipole moments. The essential thing from our point of view is that, whenever dipoles are lined up parallel to each other, they exert a mutual depolarizing effect, which must be taken into account.

The variation of the heat of adsorption with the fraction of the surface covered was first considered by Herzfield (1929), who explained the decrease in the heat of adsorption with the amount of gas adsorbed in terms of the electrostatic interactions between dipoles induced in the adsorbed molecules. Roberts and Orr (1938) considered the field near the cleavage plane of a body-centred ionic crystal in connexion with some experimental results of Orr (1939 a,b) on the adsorption of non-polar gases on ionic crystals. Roberts (1938) also considered the variation of the heat of adsorption with the fraction of the surface covered when vapours with a permanent dipole moment are adsorbed on conductors. None of these calculations takes account of the statistical distribution of particles on the surface, that is, it is assumed that there is a random distribution of particles on the surface. Wang (1938) has given an approximate method for treating adsorption problems, taking the statistical distribution of particles over the surface into account, when the adsorbed particles have a constant dipole moment.

For the case of a quadratic lattice, Wang considered a central site o and its four nearest neighbours, the first shell sites, 1, 2, 3, 4. All the ways of arranging from o to 5 particles on these sites were written down and the interaction energy of the particles in each such arrangement was calculated. The effect of the outer sites was taken into account by replacing them by a continuous distribution of uniform dipole moment per unit area equal to that of the discrete distribution. The continuous distribution was taken to start at a distance ρ from the central site and, for any value of θ, the fraction of sites occupied, ρ was chosen to make each first shell site an average site, that is, to make its probability of occupation θ. The numerical calculations involved in an application of the method are tedious, and a simpler method of taking into account the effects of the outer sites was developed subsequently by Miller (1940). Further, the joint effect of van der Waals and dipole interactions between adsorbed particles can be considered. We first consider the case in which the dipole moment is constant, and then show how the variation of the dipole moment with the fraction of the surface covered can be taken into account.

7·2. The Electrostatic Field

Topping (1927) has calculated the interaction energy of a lattice of dipoles and has shown that for a square array the repulsive field at any one dipole due to all the others is $9M/a^3$, where a and M are the lattice spacing and the dipole moment respectively. In the present approximation the radius ρ, at which the continuous distribution begins, is not taken as a function of θ, as in Wang's approximation, but is fixed. Its value is chosen to make the field at the central site, produced both by particles occupying the same set of first shell sites as above and by the continuous distribution, equal, when $\theta = 1$, to Topping's value. This condition requires that

$$4\frac{M}{a^3} + \frac{2\pi M}{a^2}\int_\rho^\infty \frac{dr}{r^2} = 9\frac{M}{a^3},$$

which gives

$$\rho = 1\cdot 25\,a. \qquad (7\cdot 1)$$

The electrostatic energy of interaction between the continuous distribution and a particle occupying the central site is denoted by U_0, and that between the continuous distribution and a particle occupying a first shell site is denoted by U_1. Wang determined U_0 and U_1 (or, more generally, U_ν, the energy of interaction of the continuous distribution outside the circle of radius ρ with an adsorbed particle at a distance a_ν from the centre of the circle) in the following way. Let x_ν, y_ν be the co-ordinates of a site at a distance a_ν from the centre of the circle, and x, y be the co-ordinates of a point situated in the continuous distribution. If $V(r)$ is the law of variation of the interaction energy of two particles with their distance apart, then

$$U_\nu = \iint_{x^2+y^2>\rho^2} V(\{(x-x_\nu)^2+(y-y_\nu)^2\}^{\frac12})\,K\theta\,dx\,dy, \qquad (7\cdot 2)$$

where K is the density of the sites and is equal to $1/a_1^2$ for a square lattice. In polar co-ordinates r, ϕ (ϕ measures the angle between the vectors (x, y) and (x_ν, y_ν)) this can be written

$$U_\nu = \frac{\theta}{a_1^2}\int_\rho^\infty dr \int_0^{2\pi} d\phi\, V(\{r^2 - 2ra_\nu\cos\phi + a_\nu^2\}^{\frac12}).$$

For dipole-dipole interaction

$$V(r) = \frac{M^2}{2r^3},$$

where M is the dipole moment; in this case we have

$$U_\nu = \frac{M^2\theta}{2a_1^2} \int_\rho^\infty r\,dr \int_0^{2\pi} \frac{d\phi}{(r^2 - 2ra_\nu \cos\phi + a_\nu^2)^{\frac{3}{2}}}. \qquad (7\cdot3)$$

Writing

$$\cos\phi = 2\cos^2\frac{\phi}{2} - 1, \quad \tfrac{1}{2}\phi = \psi, \quad \cos^2\psi = x \quad \text{and} \quad \kappa_\nu^2 = \frac{4ra_\nu}{(r+a_\nu)^2},$$

it follows that

$$r^2 - 2ra_\nu \cos\phi + a_\nu^2 = (r+a_\nu)^2 (1 - \kappa_\nu^2 \cos^2\phi),$$

and therefore

$$\int_0^{2\pi} \frac{d\phi}{(r^2 - 2ra_\nu \cos\phi + a_\nu^2)^{\frac{3}{2}}} = \frac{2}{(r+a_\nu)^3} \int_0^1 x^{-\frac{1}{2}} (1 - \kappa_\nu^2 x^2)^{-\frac{3}{2}}\,dx.$$

Using a well-known integral formula for the hypergeometric function (Whittaker and Watson 1927) this can be written in the form

$$\frac{2\pi}{(r+a_\nu)^3} F(\tfrac{1}{2}, \tfrac{3}{2}; 1; \kappa_\nu^2),$$

where $F(a, \beta; \gamma; x)$ is the hypergeometric function. Using the formula

$$F(a, \beta; 2a; x)$$
$$= \left[1 + \frac{1 - (1-x)^{\frac{1}{2}}}{1 + (1-x)^{\frac{1}{2}}} \right]^{2\beta} F\left(\beta, \beta - a + \tfrac{1}{2}; a + \tfrac{1}{2}; \left\{ \frac{1 - (1-x)^{\frac{1}{2}}}{1 + (1-x)^{\frac{1}{2}}} \right\}^2 \right), \quad (7\cdot4)$$

which is due to Kummer (1836), we obtain

$$U_\nu = \frac{\pi}{a_1^2} M^2\theta \int_\rho^\infty \frac{r\,dr}{(r+a_\nu)^3} \left(1 + \frac{a_\nu}{r}\right)^3 F\left(\tfrac{3}{2}, \tfrac{3}{2}; 1; \frac{a_\nu^2}{r^2}\right). \qquad (7\cdot5)$$

Since $\rho > a_\nu$, a_ν/r is always less than unity for all values of r with which we are concerned, and since the series under the integral sign is uniformly convergent for $|a_\nu/r| < 1$, we can therefore integrate $(7\cdot5)$ term by term; this gives

$$U_\nu = \frac{\pi M^2\theta}{a_1^2\rho} F\left(\tfrac{1}{2}, \tfrac{3}{2}; 1; \frac{a_\nu^2}{\rho^2}\right) = 2\pi\theta V_1 F\left(\tfrac{1}{2}, \tfrac{3}{2}; 1; \frac{a_\nu^2}{\rho^2}\right), \qquad (7\cdot6)$$

where V_1 is the dipole interaction energy between two particles adsorbed on closest neighbour sites. Again making use of formula $(7\cdot4)$ we obtain the result in a form more convenient for purposes of calculation:

$$U_\nu = 2\pi\theta V_1 \frac{a_1}{-} (1 + \sigma_\nu)^3 F(\tfrac{3}{2}, \tfrac{3}{2}; 1; \sigma_\nu^2), \qquad (7\cdot7)$$

where $\qquad \sigma_\nu = \{1 - (1 - a_\nu^2/\rho^2)^{\frac{1}{2}}\}/\{1 + (1 - a_\nu^2/\rho^2)^{\frac{1}{2}}\}.$

In particular, we obtain the results

$$\left.\begin{aligned}
U_0 &= 1 \cdot 6\pi\theta V_1, \\
U_1 &= 2\pi\theta V_1 \frac{a_1}{\rho}(1 + \sigma_1)^3 F(\tfrac{3}{2}, \tfrac{3}{2}; 1; \sigma_1^2) \\
&= 11 \cdot 35\theta V_1.
\end{aligned}\right\} \qquad (7 \cdot 8)$$

7·3. The Partition Function for the System

We introduce θ_i $(i = 0, 1, 2, 3, 4)$, each of which may be zero or unity, to denote the state of occupation of the central and first shell sites, and use the Bethe approximation to set up the grand partition function for the system. It can be written

$$\Xi = \Sigma_{\theta_i} \, \epsilon_0^{\theta_0} \, \epsilon_1^{\theta_1 + \cdots + \theta_4} \, \eta_1^{\theta_0(\theta_1 + \cdots + \theta_4)} \, \eta_2^{\theta_1\theta_2 + \theta_2\theta_3 + \theta_3\theta_4 + \theta_4\theta_1} \, \eta_3^{\theta_1\theta_3 + \theta_2\theta_4}, \quad (7 \cdot 9)$$

where

$$\left.\begin{aligned}
\epsilon_0 &= e^{\chi/kT}\left(\frac{h^2}{2\pi m kT}\right)^{\frac{3}{2}} \frac{p}{kT}\frac{v_s(T)}{b_g(T)} e^{-U_0/kT} = \xi\, e^{-U_0/kT}, \\
\epsilon_1 &= e^{\chi/kT}\left(\frac{h^2}{2\pi m kT}\right)^{\frac{3}{2}} \frac{p}{kT}\frac{v_s(T)}{b_g(T)} \zeta\, e^{-U_1/kT} = \xi\zeta\, e^{-U_1/kT}, \\
\eta_1 &= e^{-(V_1-V)/kT}, \quad \eta_2 = e^{-V_2/kT}, \quad \eta_3 = e^{-V_3/kT}.
\end{aligned}\right\} \qquad (7 \cdot 10)$$

In these formulae V_ν ergs is the interaction energy due to the electrostatic forces between two particles separated by a distance a_ν cm., $a_1, a_2, a_3, \ldots$ being the distances between nearest neighbours, next nearest neighbours, and so on. $-V$ ergs is the interaction energy due to van der Waals forces between a pair of neighbouring adsorbed particles, van der Waals interactions between more distant particles being neglected. $-\chi$ ergs is the energy of adsorption of an isolated dipole, $v_s(T)$, $b_g(T)$ are respectively the vibrational partition function for the adsorbed states and the partition function for the internal motion of the particles in the gas phase; p dyne cm.$^{-2}$ is the pressure in the gas phase and T° K. is the temperature. The factor ζ is the parameter introduced in the Bethe method to represent the effect of the outer sites and it allows for

(a) the van der Waals interactions between particles occupying first shell sites and those occupying outer sites;

(b) the difference between the electrostatic field due to the continuous distribution and that due to the discrete distribution, caused by the effect of the occupation of first shell sites on the probability of occupation of near outer sites.

Denote the number of particles occupying the central site by n_0 (which is either zero or unity) and the number occupying first shell sites by n_1; of these five particles let m_2, m_3, respectively, be the numbers of particles separated by distances a_2, a_3. The product $n_0 n_1$ is the number of particles separated by a distance a_1. Corresponding to given values of n_0, n_1, m_2, m_3 let there be $g(n_0, n_1; m_2, m_3)$ sets of values of θ_0, θ_1, θ_2, θ_3 and θ_4. Then $g(n_0, n_1; m_2, m_3)$ is the weight to be attached to the state specified by given numbers n_0, n_1, m_2, m_3. Equation (7·9) can be written in the form

$$\Xi = \sum_{\substack{n_0, n_1 \\ m_2, m_3}} g(n_0, n_1; m_2, m_3) \epsilon_0^{n_0} \epsilon_1^{m_1} \eta_1^{n_0 n_1} \eta_2^{m_2} \eta_3^{m_3}. \qquad (7·11)$$

For a square array the weights appropriate to given sets of values of n_1, m_2, m_3 are given in table 6. For each of these sets of values n_0 can take both the values zero and unity. Using equation (7·11) and table 6 we obtain for the grand partition function

$$\Xi = \epsilon_0 f_1(\eta_1 \epsilon_1, \eta_2, \eta_3) + f_1(\epsilon_1, \eta_2, \eta_3), \qquad (7·12)$$

where

$$f_1(x, \eta_2, \eta_3) \equiv \eta_2^4 \eta_3^2 x^4 + 4\eta_2^2 \eta_3 x^3 + 4\eta_2 x^2 + 2\eta_3 x^2 + 4x + 1.$$

TABLE 6

n_1	4	3	2	2	1	0
m_2	4	2	1	0	0	0
m_3	2	1	0	1	0	0
g	1	4	4	2	4	1

The average value of n_0 is given by

$$\bar{n}_0 = \frac{\epsilon_0 f_1(\eta_1 \epsilon_1, \eta_2, \eta_3)}{\epsilon_0 f_1(\eta_1 \epsilon_1, \eta_2, \eta_3) + f_1(\epsilon_1, \eta_2, \eta_3)}.$$

Since the central site is an average site we have $\bar{n}_0 = \theta$, whence

$$\frac{\theta}{1-\theta} = \frac{\epsilon_0 f_1(\eta_1 \epsilon_1, \eta_2, \eta_3)}{f_1(\epsilon_1, \eta_2, \eta_3)}. \qquad (7·13)$$

In a similar way the average value of n_1 is given by

$$\bar{n}_1 = \frac{4f_2(\eta_1 \epsilon_1, \eta_2, \eta_3) + 4f_2(\epsilon_1, \eta_2, \eta_3)}{\epsilon_0 f_1(\eta_1 \epsilon_1, \eta_2, \eta_3) + f_1(\epsilon_1, \eta_2, \eta_3)},$$

where

$$f_2(x, \eta_2, \eta_3) \equiv \eta_2^4 \eta_3^2 x^4 + 3\eta_2^2 \eta_3 x^3 + 2\eta_2 x^2 + \eta_3 x^2 + x = \tfrac{1}{4} x \frac{\partial}{\partial x} f_1(x, \eta_2, \eta_3).$$

Since the first shell sites are average sites, we have

$$\bar{n}_1 = 4\theta,$$

whence $\qquad \theta = \dfrac{f_2(\eta_1\epsilon_1, \eta_2, \eta_3) + f_2(\epsilon_1, \eta_2, \eta_3)}{\epsilon_0 f_1(\eta_1\epsilon_1, \eta_2, \eta_3) + f_1(\epsilon_1, \eta_2, \eta_3)}.$ $\qquad$ (7·14)

ϵ_0 can be eliminated between equations (7·13) and (7·14), yielding

$$\frac{\theta}{1-\theta}\left\{1 - \frac{f_2(\eta_1\epsilon_1, \eta_2, \eta_3)}{f_1(\eta_1\epsilon_1, \eta_2, \eta_3)}\right\} = \frac{f_2(\epsilon_1, \eta_2, \eta_3)}{f_1(\epsilon_1, \eta_2, \eta_3)}. \qquad (7\cdot15)$$

For a fixed dipole moment equation (7·15) determines θ immediately as a function of ϵ_1. Then equation (7·13) can be used to determine ϵ_0, which gives the adsorption isotherm.

7·4. The Variation of the Heat of Adsorption

The variation of the heat of adsorption with the fraction of the surface covered can be determined from the Clausius-Clapeyron equation

$$\left(\frac{\partial}{\partial T} \log p\right)_\theta = \frac{q + kT}{kT^2}. \qquad (7\cdot16)$$

Now $\qquad \xi = \left(\dfrac{h^2}{2\pi m}\right)^{\frac{3}{2}} \dfrac{e^{\chi/kT}}{(kT)^{\frac{5}{2}}} \dfrac{v_s(T)}{b_g(T)} p,$

and the heat of adsorption at $\theta = 0$ is

$$q_0 = \chi + \tfrac{3}{2}kT + kT^2 \frac{d}{dT} \log \frac{v_s(T)}{b_g(T)},$$

whence by differentiating $\log \xi$ with respect to T we obtain eventually

$$q - q_0 = kT^2 \left(\frac{\partial}{\partial T} \log \xi\right)_\theta. \qquad (7\cdot17)$$

Since $\epsilon_0 = \xi e^{-U_0/kT}$, equation (7·13) gives

$$\log \xi = \log \frac{\theta}{1-\theta} + \frac{U_0}{kT} + \log f_1(\epsilon_1, \eta_2, \eta_3) - \log f_1(\eta_1\epsilon_1, \eta_2, \eta_3).$$

Remembering that $\epsilon_1, \eta_1, \eta_2$ and η_3 are all functions of T and that

$U_0 = 1 \cdot 6\pi\theta V_1$, we obtain for the heat of adsorption due to the interaction energy of the particles

$$(q - q_0)_i = - 1 \cdot 6\pi\theta V_1$$
$$+ \left\{ \frac{\partial \epsilon_1}{\partial T} \frac{\partial}{\partial \epsilon_1} + (V_1 - V) \eta_1 \frac{\partial}{\partial \eta_1} + V_2 \eta_2 \frac{\partial}{\partial \eta_2} + V_3 \eta_3 \frac{\partial}{\partial \eta_3} \right\} \log \frac{f_1(\epsilon_1, \eta_2, \eta_3)}{f_1(\eta_1\epsilon_1, \eta_2, \eta_3)}.$$

$$(7 \cdot 18)$$

Before this equation can be used to calculate the heat of adsorption, $\partial\epsilon_1/\partial T$ must be evaluated for the required values of θ. Differentiating both members of equation $(7 \cdot 15)$ with respect to T we obtain finally

$$\frac{\partial \epsilon_1}{\partial T} \Bigg[\left\{ 1 - \frac{f_2(\eta_1\epsilon_1, \eta_2, \eta_3)}{f_1(\eta_1\epsilon_1, \eta_2, \eta_3)} \right\} \frac{\partial}{\partial \epsilon_1} \frac{f_1(\epsilon_1, \eta_2, \eta_3)}{f_2(\epsilon_1, \eta_2, \eta_3)}$$
$$- \frac{f_1(\epsilon_1, \eta_2, \eta_3)}{f_2(\epsilon_1, \eta_2, \eta_3)} \frac{\partial}{\partial \epsilon_1} \frac{f_2(\eta_1\epsilon_1, \eta_2, \eta_3)}{f_1(\eta_1\epsilon_1, \eta_2, \eta_3)} \Bigg]$$
$$= \frac{V_1 - V}{kT^2} \eta_1 \left[\frac{f_1(\epsilon_1, \eta_2, \eta_3)}{f_2(\epsilon_1, \eta_2, \eta_3)} \frac{\partial}{\partial \eta_1} \frac{f_2(\eta_1\epsilon_1, \eta_2, \eta_3)}{f_1(\eta_1\epsilon_1, \eta_2, \eta_3)} \right]$$
$$+ \frac{V_2}{kT^2} \eta_2 \left[\frac{f_1(\epsilon_1, \eta_2, \eta_3)}{f_2(\epsilon_1, \eta_2, \eta_3)} \frac{\partial}{\partial \eta_2} \frac{f_2(\eta_1\epsilon_1, \eta_2, \eta_3)}{f_1(\eta_1\epsilon_1, \eta_2, \eta_3)} \right.$$
$$\left. - \left\{ 1 - \frac{f_2(\eta_1\epsilon_1, \eta_2, \eta_3)}{f_1(\eta_1\epsilon_1, \eta_2, \eta_3)} \right\} \frac{\partial}{\partial \eta_2} \frac{f_1(\epsilon_1, \eta_2, \eta_3)}{f_2(\epsilon_1, \eta_2, \eta_3)} \right]$$
$$+ \frac{V_3}{kT^2} \eta_3 \left[\frac{f_1(\epsilon_1, \eta_2, \eta_3)}{f_2(\epsilon_1, \eta_2, \eta_3)} \frac{\partial}{\partial \eta_3} \frac{f_2(\eta_1\epsilon_1, \eta_2, \eta_3)}{f_1(\eta_1\epsilon_1, \eta_2, \eta_3)} \right.$$
$$\left. - \left\{ 1 - \frac{f_2(\eta_1\epsilon_1, \eta_2, \eta_3)}{f_1(\eta_1\epsilon_1, \eta_2, \eta_3)} \right\} \frac{\partial}{\partial \eta_3} \frac{f_1(\epsilon_1, \eta_2, \eta_3)}{f_2(\epsilon_1, \eta_2, \eta_3)} \right], \quad (7 \cdot 19)$$

in which the three terms in square brackets on the right-hand side arise from the differentiation of the terms involving η_1, η_2 and η_3 respectively with respect to T. Equations $(7 \cdot 15)$, $(7 \cdot 18)$ and $(7 \cdot 19)$ determine the variation of the heat of adsorption with the fraction of the surface covered when the dipole moment is independent of this latter quantity.

We now consider the variation of the dipole moment with the fraction of the surface covered (Miller 1941, 1946). When a fraction θ of the total number of sites is occupied, let the average dipole moment per adsorbed particle be M_θ. This dipole moment is given by

$$M_\theta = M_0 + \sigma Z_\theta, \quad (7 \cdot 20)$$

where M_0 is the permanent dipole moment, that is, the dipole moment of an isolated particle, σ is the polarization coefficient and Z_θ is the electrostatic field at any site due to all the adsorbed particles of dipole moment M_θ. The result due to Topping (1927) to which reference has already been made shows that for a complete monolayer the relation between the dipole moment and the field is

$$Z_1 = -9M_1/a_1^3. \qquad (7\cdot21)$$

When a fraction θ of the sites are occupied, so that the dipole moment of each adsorbed particle is M_θ, we assume that the electrostatic field at any site due to all the adsorbed particles is given by

$$Z_\theta = -9\theta M_\theta/a_1^3, \qquad (7\cdot22)$$

where the factor θ is introduced, since only this fraction of the total number of sites is occupied and the factor 9 is the result of summing over all the sites. From equations $(7\cdot20)$ and $(7\cdot22)$ the dipole moment M_θ can be obtained as a function of the fraction of the surface covered; it is given by

$$M_\theta = \frac{M_0}{1 + 9\sigma\theta/a_1^3}. \qquad (7\cdot23)$$

Substituting this value in equation $(7\cdot22)$ we get

$$Z_\theta = -\frac{9M_0}{a_1^3}\frac{\theta}{1 + 9\sigma\theta/a_1^3}. \qquad (7\cdot24)$$

Equations $(7\cdot18)$ and $(7\cdot19)$ can be used to calculate a series of heat curves corresponding to constant dipole moments $M_0, M_{\theta_1}, \ldots, M_{\theta_r}, \ldots, M_1$ determined by equation $(7\cdot23)$. Since the differentiation in equation $(7\cdot17)$ is to be carried out for a fixed value of θ, the point in which the ordinate at $\theta = \theta_r$ meets the M_{θ_r} member of the family of heat curves gives the contribution, at this value of θ, to the heat of adsorption *due to the interactions* between the adsorbed particles when the variation of the dipole moment with the fraction of the surface covered is taken into account. In practice the calculations are carried out in the following manner. The dipole moment is calculated from equation $(7\cdot23)$ for a particular value of θ; this determines η_1, η_2 and η_3, so that the coefficients of the terms in the functions $f_1(x, \eta_2, \eta_3)$ and $f_2(x, \eta_2, \eta_3)$ are determined for this value of θ. Then equation $(7\cdot15)$ can be solved numerically by a method of successive approximations* to give the value of ϵ_1 for this value of θ. If these corresponding values of ϵ_1 and θ are used in equations (7·18) and (7·19), the heat of adsorption (denoted by $(q-q_0)_i$) which arises from the interaction energy of the adsorbed particles for this value of θ, when the variation of the dipole moment with the fraction of the surface covered is taken into account, can be determined.

It is clear that, when a fraction θ of the surface is covered, this process gives the contribution to the heat of adsorption, which corresponds to the change in the interaction energy following the adsorption of an additional particle of dipole moment M_θ. But, in the actual adsorption process, a particle of dipole moment M_0 must first have its dipole moment reduced to M_θ. The work done on the particle due to this depolarization is given by $\Delta U = (M_0 - M_\theta)^2/2\sigma$, and, as this is the change in energy which takes place during the adsorption of a single particle, there is a corresponding contribution to the heat of adsorption given by

$$(q-q_0)_p = -(M_0 - M_\theta)^2/2\sigma. \qquad (7·25)$$

This particle of dipole moment M_θ can then be brought to the surface and adsorbed. The adsorption of this particle changes the interaction energy due to the van der Waals and to the electrostatic forces; corresponding to this change in the interaction energy of the adsorbed layer there is a further contribution, which has been denoted by $(q-q_0)_i$ to the heat of adsorption. $(q-q_0)_i$ is the quantity which is calculated in the way indicated above from equations (7·18) and (7·19) with the appropriate values of the dipole moment and of η_1, η_2 and η_3 for each value of θ. The total variation of the heat of adsorption when the variation of the dipole moment with the fraction of the surface covered is taken into account is given by the sum of $(q-q_0)_i$ and $(q-q_0)_p$.

7·5. Numerical Calculations

These results will be applied to consider the joint effect of van der Waals forces and the dipole-dipole interactions. The numerical calculations are carried out for the adsorption of ammonia on a non-metallic crystal which has no net electrostatic charge.

* If the $\epsilon_1 - \theta$ relation corresponding to a fixed dipole moment, not too greatly different from M_θ, is known, this numerical solution is fairly easy to carry out.

The van der Waals interaction is assumed to be one-sixth of the latent heat of vaporization per molecule,

$$-V = -6\cdot4 \times 10^{-14} \text{ erg.}$$

The dipole moment is taken from the Faraday Society Tables (1934),

$$M_0 = 1\cdot5 \times 10^{-18} \text{ e.s.u.}$$

The lattice spacing which depends on the adsorbing surface is taken as

$$a = 4 \times 10^{-8} \text{ cm.}$$

These values give

$$V_1 = 3\cdot52 \times 10^{-14} \text{ erg,} \quad V_2 = \sqrt{2}\,V_1/4, \quad V_3 = V_1/8.$$

The polarization coefficient can be calculated, by means of the Clausius-Mosotti formula, from some experimental measurements by Zahn (1926) of the dielectric constant of the ammonia molecule. For it we obtain

$$\sigma = 2\cdot3 \times 10^{-24}.$$

The calculations are first carried out for a fixed dipole moment equal to M_0, using equations (7·15), (7·18) and (7·19). As $\epsilon_1 \to \infty$ or $\theta \to 1$ the third term of the right-hand member of equation (7·18) has the limiting value $(V_1 - V)/V_1$, since

$$\frac{\partial}{\partial \eta_1} \log \frac{f_1(\epsilon_1, \eta_2, \eta_3)}{f_1(\eta_1\epsilon_1, \eta_2, \eta_3)} \sim -\frac{4}{\eta_1}$$

as $\epsilon_1 \to \infty$. In the case considered here the ratio of the van der Waals to the electrostatic interactions between closest neighbours is equal to 3·636, so that for $\theta = 1$, $(q - q_0)/zV_1 = -0\cdot439$. If Q calories is the heat of adsorption per gram-molecule then this corresponds to $Q_1 - Q_0 = -882$ which is very much less than when van der Waals forces alone are operative. In fig. 32 the variation of $Q - Q_0$ with θ when van der Waals forces alone are operative is represented by the upper curve, while the variation of $Q - Q_0$ with θ when both van der Waals and electrostatic forces are operative is represented by the lower curve. The former is calculated from equation (2·16). Thus the direct electrostatic forces considerably reduce the variation of the heat of adsorption. This is in agreement with the result obtained by Roberts (1938c) using a method which disregarded the statistical distribution of particles over the surface. The effects of the van der Waals and the dipole interactions are opposite in sign and of the same order of magnitude, so that the resultant variation of $Q - Q_0$ with θ is very much less than would be expected from a consideration of forces of one type only.

The effect of the variation of the dipole moment with the fraction of the surface covered will now be taken into account.

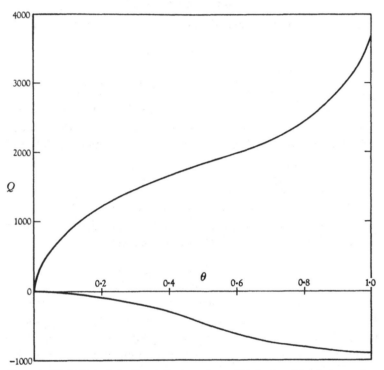

Fig. 32. Variation of the heat of adsorption with the fraction of the surface covered. The lower curve shows the effect of both electrostatic and van der Waals forces, and the upper curve is that of van der Waals forces alone.

The values of the interaction energies and of the coefficients η_1, η_2 and η_3 at 290° K. and for different values of θ are given in table 7. For each value of θ the corresponding value of ϵ_1 has been determined from equation (7·15), and the heat of adsorption which arises from the electrostatic and van der Waals interaction energy has been determined from equations (7·18) and (7·19) using the appropriate values of η_1, η_2 and η_3 given in table 7. The contribution to the heat of adsorption which arises from the mutual depolarizing action of the molecules has been determined from equation (7·25). The values which are obtained are shown in table 8, in which the unit is calories per gram-molecule. Curve (a) of fig. 33 shows the variation of the heat of adsorption of ammonia with the fraction of the surface covered when the variation of the dipole moment with the fraction of the surface covered and the statistical distribution of the particles over the surface are taken into account. Curve (b) shows the

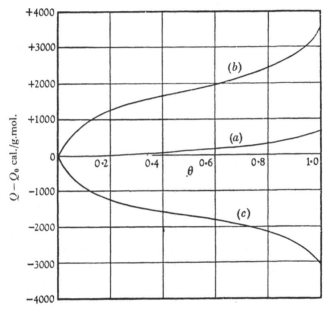

Fig. 33. Variation of the heat of adsorption when the statistical distribution of the adsorbed particles over the surface and the variation of the dipole moment are taken into account. Curve (a) shows the resultant variation due to van der Waals and electrostatic forces, curve (b) shows that due to van der Waals forces alone, and curve (c) shows that due to the electrostatic forces alone.

variation of the heat of adsorption due to the van der Waals forces alone; it has been calculated from equation (2·16). Curve (c) shows the variation of the heat of adsorption due to electrostatic forces alone when the variation of the dipole moment with the fraction of the surface covered is taken into account. From these curves we see that the contributions to the heat of adsorption due to the van der Waals forces and to electrostatic forces are of the same order of magnitude and opposite in sign, so that the resultant variation is very much smaller than when forces of only one kind are operative.

TABLE 7

θ	V_1 10^{14} erg	V_2 10^{14} erg	V_3 10^{14} erg	η_1	η_2	η_3
0	3·52	1·24	0·439	2·05	0·733	0·896
0·2	3·10	1·10	0·388	2·28	0·760	0·908
0·4	2·76	0·974	0·345	2·49	0·784	0·917
0·6	2·47	0·872	0·308	2·67	0·804	0·925
0·8	2·22	0·784	0·277	2·84	0·822	0·932
1·0	2·01	0·710	0·251	3·00	0·837	0·939

TABLE 8

θ	0·2	0·4	0·6	0·8	1·0
$(Q-Q_0)_i$	24·5	173·9	359·0	593·9	1077·7
$(Q-Q_0)_p$	−26·1	−92·8	−186·8	−298·7	−422·2
Total	− 1·6	+81·1	+172·2	+295·2	+655·5

In fig. 34 the variation of that part of the heat of adsorption which arises from the electrostatic forces is shown as a function of the fraction of the surface covered. In curve (a) the variation of the dipole moment with the fraction of the surface covered is taken into account, while curve (b) is calculated for a fixed dipole moment equal to M_0 and curve (c) is calculated for a fixed dipole moment equal to M_1. The dotted curve shows that part of the heat of adsorption which is contributed by the *electrostatic interactions* between the adsorbed particles when the variation of the dipole moment is taken into account, that is, the total contribution to the heat of adsorption of electrostatic origin less that part which arises from the mutual depolarizing action of the adsorbed particles. This curve is, of course, tangential to curve (b) at $\theta = 0$ and to curve (c) at $\theta = 1$. The difference between curve (a) and the dotted curve at each point is the contribution to the heat of adsorption due to the mutual depolarizing effect of the adsorbed particles, and, in particular, it is on account of this effect that the end-point (at $\theta = 1$) of curve (a) does not coincide with that of curve (c).

Curve (*a*) of fig. 35 shows the variation of the heat of adsorption due to the electrostatic and the van der Waals forces together, when the variation of the dipole moment with the fraction of the surface covered is taken into account, and curve (*b*) shows the

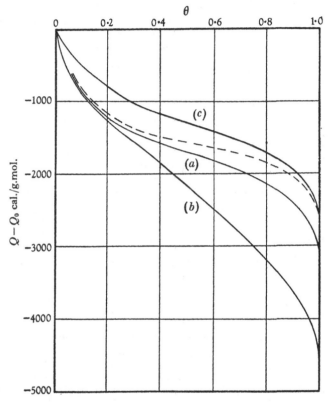

Fig. 34. Variation of the contribution to the heat of adsorption due to the electrostatic forces. Curve (*a*) takes into account the variation of the dipole moment, curve (*b*) is calculated for a fixed dipole moment equal to M_0, curve (*c*) is calculated for a fixed dipole moment equal to M_1, and the broken curve shows the contribution which is due to the electrostatic interactions.

corresponding quantity calculated for a fixed dipole moment equal to M_0. The difference between these two curves is accounted for by the difference between the contributions to the heat of adsorption which arise from the electrostatic forces in the two cases (and shown by curves (*a*) and (*b*) of fig. 34). Both curves show the small resultant variation in the heat of adsorption when both the van der Waals and electrostatic forces are operative compared with that due to either type of force alone. Whether or not the resultant value of $Q - Q_0$ is positive or negative depends upon which of the two kinds of force predominates.

The effect of the statistical distribution of the particles on the surface can be determined by deriving an expression for the energy

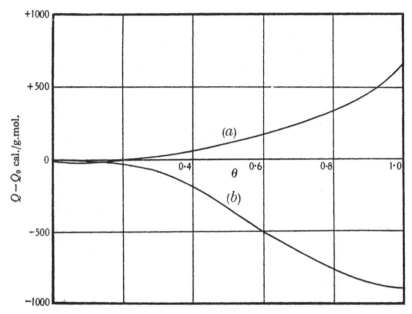

Fig. 35. Resultant variation of the heat of adsorption showing the effect of the variation of the dipole moment. Curve (a) takes account of the variation of the dipole moment, and curve (b) is for a fixed dipole moment equal to M_0 Both curves take the statistical distribution of the adsorbed particles over the surface into account.

of a film which consists of a random distribution of particles on a non-conducting surface, when the variation of the dipole moment with the fraction of the surface covered is taken into account. Consider a large number of similar particles, each of which has a dipole moment equal to M_0, and which are all a large distance apart. While they are outside the influence of one another, change the dipole moment of each so that it becomes equal to M_θ. We have seen that the work done on each dipole is equal to

$$(M_0 - M_\theta)^2 / 2\sigma,$$

where M_θ is given by equation (7·23). Now move the dipoles to the surface of a non-conducting solid so that they form a random distribution on a square lattice and occupy a fraction θ of the total number of lattice sites. The work done on each dipole during this process is

$$-\tfrac{1}{2}M_\theta Z_\theta, \qquad (7\cdot26)$$

where Z_θ is given by equation (7·24) and the factor one-half is introduced so that each interaction is included once only. It follows that the energy U of the film due to dipole interactions is given by

$$U = -\tfrac{1}{2}N_s\theta M_\theta Z_\theta + \tfrac{1}{2}N_s\theta\sigma Z_\theta^2,$$

in which the factor θ is introduced, since that fraction of the total number of sites on the surface is occupied. Using equation (7·20) this can be written in the form

$$U = -\tfrac{1}{2}N_s\theta M_0 Z_\theta. \qquad (7\cdot27)$$

If we eliminate Z_θ between equations (7·24) and (7·27) the interaction energy of the film can be written as

$$U = \frac{9N_s M_0^2}{2a_1^3}\frac{\theta^2}{1+9\sigma\theta/a_1^3}. \qquad (7\cdot28)$$

Therefore, for a random distribution of particles over the surface, the heat of adsorption per particle arising from the electrostatic forces is given by

$$(q-q_0)_{\text{elec.}} = -\frac{\partial U}{\partial(N_s\theta)} = -\frac{9M_0^2}{2a_1^3}\frac{\theta(2+9\sigma\,\theta/a_1^3)}{(1+9\sigma\theta/a_1^3)^2}. \qquad (7\cdot29)$$

For a random distribution of particles, the energy of the film due to the van der Waals interactions between particles adsorbed on closest neighbour sites is given by

$$U_w = -\tfrac{1}{2}N_s z V\theta^2, \qquad (7\cdot30)$$

and the contribution $(q-q_0)_w$ to the heat of adsorption per particle which arises from these forces is given by

$$(q-q_0)_w = z V\theta. \qquad (7\cdot31)$$

The resultant variation of the heat of adsorption per particle for a random distribution of particles on the surface is the sum of $(q-q_0)_{\text{elec.}}$ and $(q-q_0)_w$.

This resultant is plotted as curve (b) in fig. 36, and the curve which is obtained when the statistical distribution of the particles over the surface is taken into account is plotted as curve (*a*). It is important to notice that the total variation in the heat of adsorption from $\theta = 0$ to $\theta = 1$ is the same in the two cases, for both methods give the total change in the heat of adsorption exactly. These curves show the effect of the clustering of the particles on the variation of the heat of adsorption.

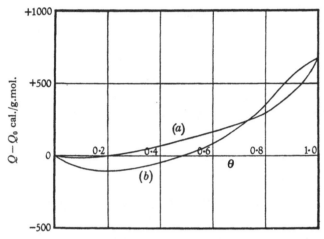

Fig. 36. Effect of the statistical distribution of the adsorbed particles over the surface on the resultant heat of adsorption. Both curves take the variation of the dipole moment of the adsorbed particles into account; curve (*a*) shows the effect of the statistical distribution of particles over the surface, while curve (*b*) is for a random distribution of particles.

The essential point is that the total variation in the heat of adsorption is much less than would be expected from a consideration of the van der Waals forces alone. This is in agreement with the experimental results of Orr (1939 *a*, *b*) on the adsorption of argon by caesium iodide. Similar results have been found by Williams (1917) for the adsorption of sulphur dioxide on charcoal. After an initial rapid drop, which is always found in experiments with charcoal,* the variation in the heat of adsorption is quite small until bulk condensation begins. At 400 cm.³ per g. the adsorption isotherm bends rapidly and, as can be seen from fig. 37, the heat is the normal latent heat so that bulk condensation has started by this stage. The important point is that between the

* Cf. Barrer (1937).

initial drop to about 9400 cal. and the onset of effects due to the formation of multilayers there is only a comparatively slight variation of the heat of adsorption in the agreement with the theory which has been developed above.

More recently, Crawford and Tompkins (1948) have carried out an experimental investigation of a number of polar and non-polar gases on barium fluoride crystals. The gases used were sulphur dioxide, ammonia, carbon dioxide and nitrous oxide. By considering the potential field provided by the surface they present

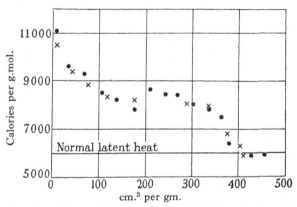

Fig. 37. Experimental values of the variation of the heat of adsorption of SO₂ on charcoal with the amount adsorbed.

evidence to show that, for the first adsorbed layer and over the range of temperatures at which measurements were carried out, these gases formed an immobile film on an array of sites determined by the geometry of the underlying crystal surface. Their results are shown in fig. 38, in which the heat of adsorption is plotted as a function of the volume of gas adsorbed. The almost linear variation of the heat of adsorption with the fraction of the surface covered in the case of the non-polar molecules provides additional evidence that the molecules are adsorbed as an immobile film. It is, however, difficult to see why the behaviour of the non-polar gases should be no different from that of the polar gases. In fact, equations (7·29) and (7·31) indicate that in the latter case the mutual depolarizing action of the adsorbed molecules should cause both an appreciable departure from the linear variation of the heat of adsorption with the fraction of the surface covered and also a reduction in the total change in the heat of adsorption. It seems likely that a more precise method than any so far developed is needed to examine the effect of long-range forces such as dipole-dipole interactions.

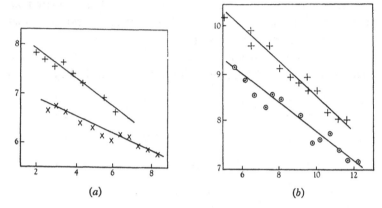

Fig. 38. Experimental values of the heat of adsorption of nitrous oxide ($\times$), carbon dioxide ($+$), ammonia ($\odot$) and sulphur dioxide ($+$) on barium fluoride crystals with the amount of gas adsorbed.

7·6. The Treatment of Long-range Forces

In the theoretical treatment of the problems which have been examined in this monograph, use has been made of the Bethe approximation. In the study of the similar statistical problem which arises in considering various second-order transitions, Kramers and Wannier (1941; Wannier 1945) and Onsager (1944, 1947) have been able to give an exact treatment, but so far only for one- and two-dimensional models and for nearest neighbour interactions. Recently, Domb (1948), using a variant of their methods, has been able formally to include a three-dimensional model in his treatment. Essentially, the solution is expressed formally as an infinite matrix having a characteristic structure (a duo-diagonal matrix). Further progress on these lines depends upon putting this solution into a tractable form. Its importance in the present connexion, however, is that by this method, it would be possible to take account of other than nearest neighbour interactions. That is, interactions between particles at any distance apart could be taken into account explicitly, and the solution would be given formally as the same type of infinite matrix. Provided a method of evaluating these matrices can be found, this would provide an exact method of dealing with long-range forces, such as the electrostatic interactions which have been considered in this chapter.

REFERENCES

BARRER, R. M. (1937). *Proc. Roy. Soc.* A, **161**, 476.

BETHE, H. A. (1935). *Proc. Roy. Soc.* A, **150**, 552.

BLODGETT, K. B. and LANGMUIR, I. (1932). *Phys. Rev.* **40**, 78.

BLÜH, O. and STARK, N. (1927). *Z. Phys.* **43**, 575.

BONHOEFFER, K. F. and FARKAS, A. (1931). *Z. phys. Chem.* B, **12**, 331.

BOSWORTH, R. C. L. (1937). *Proc. Camb. Phil. Soc.* **33**, 394.

BOSWORTH, R. C. L. (1945 a). *J. Roy. Soc. N.S.W.* **79**, 53.

BOSWORTH, R. C. L. (1945 b). *J. Roy. Soc. N.S.W.* **79**, 166.

BOSWORTH, R. C. L. and RIDEAL, E. K. (1937). *Proc. Roy. Soc.* A, **162**, 1.

BRADLEY, R. S. (1936). *J. Chem. Soc.* p. 1799.

BRYCE, G. (1936). *Proc. Camb. Phil. Soc.* **32**, 648.

BURDON, R. S. (1935). *Proc. Phys. Soc.* **47**, 460.

BURDON, R. S. (1940). *Surface Tension and Spreading of Liquids*, p. 24. Cambridge University Press.

CHANG, T. S. (1938). *Proc. Camb. Phil. Soc.* **34**, 224.

DE BOER, J. H. (1935). *Electron Emission and Adsorption Phenomena.* Cambridge University Press.

DE BOER, J. H. and ZWICKER, C. (1929). *Z. phys. Chem.* B, **3**, 407.

DEVONSHIRE, A. F. (1937). *Proc. Roy. Soc.* A, **158**, 269.

DOMB, C. (1949). *Proc. Roy. Soc.* A, **196**, 36.

FOWLER, R. H. (1935). *Proc. Camb. Phil. Soc.* **31**, 260.

FOWLER, R. H. (1936). *Proc. Camb. Phil. Soc.* **32**, 144.

FOWLER, R. H. and GUGGENHEIM, E. A. (1939). *Statistical Thermodynamics*, Ch. x. Cambridge University Press.

FRANKENBURG, W. G. (1944). *J. Amer. Chem. Soc.* **66**, 1827.

FRANKENBURGER, W. and HODLER, A. (1932). *Trans. Faraday Soc.* **28**, 229.

GIAUQUE, W. F. (1930). *J. Amer. Chem. Soc.* **52**, 4816.

GUGGENHEIM, E. A. (1938). *Proc. Roy. Soc.* A, **169**, 134.

HALSEY, G. and TAYLOR, H. S. (1947). *J. Chem. Phys.* **15**, 624.

HERZFELD, K. F. (1929). *J. Amer. Chem. Soc.* **51**, 2608.

JACKSON, J. M. (1932). *Proc. Camb. Phil. Soc.* **28**, 136.

JACKSON, J. M. and HOWARTH, A. (1933). *Proc. Roy. Soc.* A, **142**, 447.

JACKSON, J. M. and HOWARTH, A. (1935). *Proc. Roy. Soc.* A, **152**, 515.

JACKSON, J. M. and MOTT, N. F. (1932). *Proc. Roy. Soc.* A, **137**, 703.

JEANS, J. H. (1925). *Dynamical Theory of Gases.* Cambridge University Press.

JOHNSON, R. P. (1938). *Phys. Rev.* **54**, 459.

JOHNSON, M. C. and HENSON, A. F. (1938). *Proc. Roy. Soc.* A, **165**, 148.

JOHNSON, M. C. and VICK, F. A. (1935). *Proc. Roy. Soc.* A, **151**, 308.

KNUDSEN, M. (1911). *Ann. Phys., Lpz.*, **34**, 593.

KNUDSEN, M. (1915). *Ann. Phys., Lpz.*, **46**, 641.

KRAMERS, H. A. and WANNIER, G. N. (1941). *Phys. Rev.* **60**, 252, 263.

KÜMMER, E. E. (1836). *J. reine angew. Math.* **15**, 77.

LANDAU, L. (1935). *Phys. Z. Sowjet.* **8**, 489.

LANGMUIR, I. (1912). *J. Amer. Chem. Soc.* **34**, 1310.

LANGMUIR, I. (1915). *J. Amer. Chem. Soc.* **37**, 417.

LANGMUIR, I. (1926). *Gen. Elect. Rev.* **29**, 153.
LANGMUIR, I. (1932). *J. Amer. Chem. Soc.* **54**, 2798.
LANGMUIR, I. and KINGDON, K. H. (1919). *Phys. Rev.* **34**, 129.
LANGMUIR, I. and KINGDON, K. H. (1925). *Proc. Roy. Soc.* A, **107**, 61.
LANGMUIR, I. and TAYLOR, J. B. (1933). *Phys. Rev.* **44**, 423.
LANGMUIR, I. and VILLARS, D. S. (1931). *J. Amer. Chem. Soc.* **53**, 486.
LENEL, F. V. (1933). *Z. phys. Chem.* B, **23**, 379.
LENNARD-JONES, J. E. (1932). *Trans. Faraday Soc.* **28**, 333.
LENNARD-JONES, J. E. and DENT, B. M. (1928). *Trans. Faraday Soc.* **24**, 92.
LENNARD-JONES, J. E. and DEVONSHIRE, A. F. (1936). *Proc. Roy. Soc.* A, **156**, 6.
LENNARD-JONES, J. E. and GOODWIN, E. T. (1937). *Proc. Roy. Soc.* A, **163**, 101.
MICHELS, W. C. (1937). *Phys. Rev.* **52**, 1067.
MILLER, A. R. (1940). *Proc. Camb. Phil. Soc.* **36**, 69.
MILLER, A. R. (1941). Thesis, Cambridge, § 4.8.
MILLER, A. R. (1946). *Proc. Camb. Phil. Soc.* **42**, 292.
MILLER, A. R. (1947). *Proc. Camb. Phil. Soc.* **43**, 232.
MILLER, A. R. (1948a). *J. Chem. Phys.* **16**, 841.
MILLER, A. R. (1948b). *The Theory of Solutions of High Polymers*, pp. 98–102. Clarendon Press, Oxford.
MILLER, A. R. and ROBERTS, J. K. (1941). *Proc. Camb. Phil. Soc.* **37**, 82.
MORRISON, J. L. and ROBERTS, J. K. (1939). *Proc. Roy. Soc.* A, **173**, 1.
ONSAGER, L. (1944). *Phys. Rev.* **65**, 117.
ONSAGER, L. (1947). *Report of Physical Society Conference at Cambridge*, July 1946, **2**, 137.
ORR, W. J. C. (1939a). *Proc. Roy. Soc.* A, **173**, 349.
ORR, W. J. C. (1939b). *Trans. Faraday Soc.* **35**, 1247.
PEIERLS, R. (1936). *Proc. Camb. Phil. Soc.* **32**, 471.
REIMANN, A. L. (1935). *Phil. Mag.* **20**, 594.
ROBERTS, J. K. (1930). *Proc. Roy. Soc.* A, **129**, 152.
ROBERTS, J. K. (1932a). *Proc. Roy. Soc.* A, **135**, 192.
ROBERTS, J. K. (1932b). *Trans. Faraday Soc.* **28**, 395.
ROBERTS, J. K. (1935a). *Proc. Roy. Soc.* A, **152**, 445.
ROBERTS, J. K. (1935b). *Proc. Roy. Soc.* A, **152**, 473.
ROBERTS, J. K. (1935c). *Nature, Lond.*, **135**, 1037.
ROBERTS, J. K. (1936). *Proc. Camb. Phil. Soc.* **32**, 152.
ROBERTS, J. K. (1937). *Proc. Roy. Soc.* A, **161**, 141.
ROBERTS, J. K. (1938a). *Proc. Camb. Phil. Soc.* **34**, 399.
ROBERTS, J. K. (1938b). *Proc. Camb. Phil. Soc.* **34**, 577.
ROBERTS, J. K. (1938c). *Trans. Faraday Soc.* **34**, 1342.
ROBERTS, J. K. (1940). *Proc. Camb. Phil. Soc.* **36**, 53.
ROBERTS, J. K. and BRYCE, G. (1936). *Proc. Camb. Phil. Soc.* **32**, 653.
ROBERTS, J. K. and MILLER, A. R. (1939). *Proc. Camb. Phil. Soc.* **35**, 293.
ROBERTS, J. K. and ORR, W. J. C. (1938). *Trans. Faraday Soc.* **34**, 1346.
ROWLEY, H. H. and BONHOEFFER, K. F. (1933). *Z. phys. Chem.* B, **21**, 84.
RUSHBROOKE, G. S. (1938). *Proc. Camb. Phil. Soc.* **34**, 424.
SODDY, F. and BERRY, A. J. (1911). *Proc. Roy. Soc.* A, **84**, 576.

STRACHAN, C. (1937). *Proc. Roy. Soc.* A, **158**, 191.

TAYLOR, J. B. and LANGMUIR, I. (1933). *Phys. Rev.* **44**, 423.

TOMPKINS, F. C. and CRAWFORD, V. A. (1948). *Trans. Faraday Soc.* **44**, 698.

TOPPING, J. (1927). *Proc. Roy. Soc.* A, **114**, 67.

VAN CLEAVE, A. B. (1938). *Trans. Faraday Soc.* **34**, 1174.

WANG, J.-S. (1937). *Proc. Roy. Soc.* A, **161**, 127.

WANG, J.-S. (1938). *Proc. Camb. Phil. Soc.* **34**, 238.

WANNIER, G. H. (1945). *Rev. Mod. Phys.* **17**, 50.

WHITTAKER, E. T. and WATSON, G. N. (1927). *Modern Analysis*, 4th ed. Cambridge University Press.

WILKINS, F. J. (1938). *Proc. Roy. Soc.* A, **164**, 496.

WILLIAMS, A. M. (1917). *Proc. Roy. Soc. Edinb.* **37**, 161.

ZAHN, C. T. (1926). *Phys. Rev.* **27**, 455.

INDEX

Printed in the United States
By Bookmasters